PRINCIPLES OF RADIATION CHEMISTRY

PRINCIPLES OF RADIATION CHEMISTRY

JAMES H. O'DONNELL, B.Sc.(Hons.) W.Aust., Ph.D. Leeds

Senior Lecturer, Department of Chemistry, University of Queensland, Brisbane, Australia

and

DAVID F. SANGSTER, B.Sc.(Hons.) Adelaide, A.R.A.C.I.

Senior Principal Research Scientist, Australian Atomic Energy Commission Research Establishment, Sutherland, New South Wales, Australia

AMERICAN ELSEVIER PUBLISHING COMPANY, INC.

NEW YORK

First published 1970

American Elsevier Publishing Company, Inc.
52 Vanderbilt Avenue,
New York, N.Y. 10017

First published in Great Britain by
Edward Arnold (Publishers) Ltd.

SBN 444 19665 X

Printed in Great Britain by Richard Clay (The Chaucer Press), Ltd.,
Bungay, Suffolk

Preface

This book was written because of the need for a brief and simple introduction to radiation chemistry. It is intended for readers with some knowledge of chemistry, who wish to make their first acquaintance with radiation chemistry. In particular, it should be suitable for undergraduate and graduate students.

This is an age when the boundaries between the main branches of chemistry are becoming more diffuse, and studies are becoming interdisciplinary within chemistry and within science. We have found that a course in radiation chemistry clearly illustrates this trend. Radiation chemistry is far from being a narrow specialization. It draws upon, and contributes to, most other fields of chemistry and branches of science.

The study of the effects of radiation on chemical systems provides information about the mechanisms and rates of chemical reactions, particularly those involving free radicals and ions, which are not readily investigated by conventional methods. For these reasons, many undergraduate and graduate courses now include a series of lectures on radiation chemistry.

This book should be useful to scientific workers in fields such as radiation biology and nuclear engineering, who wish to obtain an understanding of the chemical effects of radiation. Also, we hope that many of our colleagues, who are interested in other aspects of chemistry, will find this a readable account of another field, opening up new horizons and perhaps illuminating their own investigations.

We would like to emphasize that this is not a comprehensive, authoritative account of radiation chemistry. Following the recent, rapid advances in the subject a number of principles are now clear. Therefore, the time is appropriate for an introductory outline of the subject. There are many omissions and simplifications for the sake of brevity and clarity, but we hope that the main developments and interpretations are clear.

We do not intend this book to act as a laboratory manual, but have incorporated useful practical information not readily available elsewhere.

In general, we have adopted and described the widely accepted points of view in the field. However, we have not hesitated to point out where uncertainties in interpretation exist and topics which are still the subject of research. We hope that the reader will not be unduly dismayed by the number of unresolved questions that still exist in radiation chemistry, but rather that he may be stirred to investigate some of the problems.

In science even the most widely accepted ideas and theories are held provisionally. This is certainly true in radiation chemistry. When the limitations of these ideas and theories become apparent, they will be rejected and replaced by better and more complete concepts.

The present knowledge of radiation chemistry is the result of the contributions of many research workers. The authors would like to dedicate this volume to them.

We would like to thank those of our colleagues who have helped us by criticism of our manuscript. We are indebted to the Australian Atomic Energy Commission and the Australian Institute of Nuclear Science and Engineering for the opportunity to carry out research in the field of radiation chemistry.

J. O'D.

1969 D. F. S.

Contents

PREFACE v

1. RADIATION CHEMISTRY 1
Fundamental studies 2
Importance 2
Historical 3
Literature 4

2. INTERACTION OF RADIATION WITH MATTER 8
Types of radiation 8
Physical aspects 15
Chemical aspects 21

3. EXPERIMENTAL PROCEDURES 31
Radiation sources 31
Dosimetry 37
Safety 42
Purification of materials 44
Treatment of samples 46

4. SPECIAL TECHNIQUES 49
Competition kinetics 49
Interrupted doses 51
Flow methods 52
Electron spin resonance 53
Matrix isolation 56
Pulse radiolysis 58
Theory and calculations 62

5. GASEOUS SYSTEMS 65
Early studies 65
Ion current 65
Comparison with condensed systems 68
The primary reactions 69
Investigational techniques 72
Typical examples 74

6. AQUEOUS SOLUTIONS 80
Primary yields 81
Molecular products 82
Primary radical species 83
Concentrated solutions 94
Typical examples 95

7. ORGANIC COMPOUNDS 99
Electron escape 99
Radical and ion–molecule reactions 101
Typical examples 102
Radiation protection 110

8. MONOMERS AND POLYMERS 113
Monomers 113
Polymers 119
Graft copolymerization 124

9. RADIATION BIOLOGY AND BIOCHEMICALS 128
Radiation damage 128
Biochemicals 131

10. SOME OTHER SYSTEMS 138
Inorganic solids 138
Heterogeneous catalysis 143
Hot atom chemistry 144
Materials 145
Decomposition of labelled compounds 146

11. FURTHER DEVELOPMENTS 148
Pulse radiolysis 148
Molecular beams 148
Electron–nuclear double resonance 150
Radiolysis at high pressures 151
Theoretical calculations 151
Other fields of chemistry 151
Radiation preservation and sterilization 154
Industrial chemical processes 156

APPENDICES
1 Definitions, units and conversion factors 160
2 Typical calculations 165

INDEX 171

Chapter 1

Radiation Chemistry

Radiation chemistry is the study of the chemical effects of high energy, ionizing radiation. High energy radiation includes electromagnetic radiation (x-rays and γ-rays), particles (α-particles, β-particles or electrons, protons and neutrons) and fission fragments.

Radiochemistry, which is often confused with *radiation chemistry*, deals with the chemistry of radioactive elements and with the use of radioactive tracers and the measurement of their radioactivity. *Nuclear chemistry* is concerned with nuclear transformations, particularly fission product and transuranium elements.

The principal characteristic of high-energy radiation is that it causes ionization in all materials. This provides a distinction between radiation chemistry and *photochemistry*—the chemical effects of light, especially ultraviolet light. The boundary is not sharp, but can be taken as about the energy of the most firmly bound outer electrons, viz. 30 eV (5 aJ) per molecule, corresponding to 400 Å (40 nm) in the vacuum ultraviolet. In *photochemistry* the energy absorption is quantized, i.e. usually a quantum of light of a certain narrow range of wavelengths undergoes a resonance interaction with a molecule of a particular component of the mixture to excite only one electron and perhaps break one specific bond. However, ultraviolet light of short wavelength can ionize certain molecules. In *radiation chemistry*, only part of the radiation energy may be transferred at an interaction and the process is not selective. There is sufficient energy available to break any bond, but in practice certain bonds may be broken preferentially. The ionizing photon or particle and the displaced electron are often both capable of producing further ionization. Thus, one incident photon may affect many thousands of molecules.

Radiation chemistry is a rapidly developing field. Hardly a month goes by without a significant addition to our knowledge and understanding of the subject, and hardly a year without an important discovery or development.

FUNDAMENTAL STUDIES

Fundamental studies in radiation chemistry aim to identify the *various species* formed in particular systems and to understand the *physical processes* by which they arise. The *amounts* formed for a given dose of radiation are measured. Then the *chemistry* of the species in their reactions with each other and with other compounds present is studied. This involves investigation of reaction kinetics and mechanisms, and of intermediates through to the final, stable end-products. The general scheme of reactions is shown in Fig. 1.1.

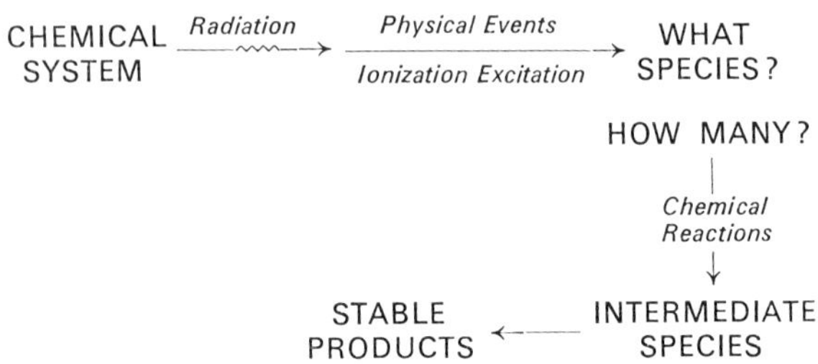

Fig. 1.1 Scheme of reactions in radiation chemistry.

IMPORTANCE

The use of radiation to initiate chemical reactions on an *industrial scale* has received much attention and the present industry is worth many millions. Processes in pilot-plant or small-scale operation now include polymerization of monomers absorbed in wood or as surface coatings on wood and metals, hydrobromination of ethylene to ethyl bromide and controlled degradation and crosslinking of polymers. There has been over-optimism in the past about the industrial development of radiation chemical processes. However, the great amount of investigational work being done is leading to many further industrial applications.

The behaviour of various materials in radiation fields is important in the construction of *nuclear reactors* and radiation sources. Much of this information has been gathered on an empirical basis, but development of fundamental knowledge of radiation chemistry can be expected to lead to improvements in materials.

Biological effects of radiation have always been of interest. Studies in this field range from physiological and biochemical effects on living whole animals, through those on tissues and other body components to the chemistry of proteins and amino acids. One aim of such work is the treatment of people inadvertently exposed to radiation. Radiation biological studies are also helping to reveal the chemistry of many life processes.

An important commercial, biological application of radiation is *bacterial sterilization*. This is widely used for surgical items, such as hypodermic syringes, and is being extended to pharmaceuticals. It has also been used for human organs, such as heart valves, being transplanted from one person to another. Other applications include disinfestation of wheat, preservation of food, and sterilization (in the reproductive sense) of insect pests. In all cases the possibility of undesirable radiation chemical side effects must not be ignored.

HISTORICAL

The first radiation chemistry effect studied was the blackening of photographic emulsions observed with x-rays by Roentgen in 1895 and with natural radioactivity by Becquerel in 1896. This was followed by the famous studies of Madame Curie, which included the discovery and isolation of polonium and radium. The chemical effects of radiation were investigated with increasing interest. Debierne studied water, Kailan organic compounds, Lind gases, Mund polymerization, while Fricke was interested in aqueous solutions. Because *ions* were observed in gases subjected to high-energy radiation, it was proposed that the chemical effects resulted from the reactions of these ions.

During the Second World War the first nuclear reactor was developed, as part of the Manhattan project, and a new era in radiation chemistry started. The availability of radioactive isotopes from nuclear reactors, the construction of accelerating machines and the needs of atomic energy developments, e.g. in reactor coolants and moderators and in the reprocessing of spent fuel elements, resulted in widespread radiation chemistry investigations.

Many who had been trained as photochemists recognized the presence of *free radicals*, and free radical reactions now became the main subject of investigation and were accepted as predominant. Most of this work was done using additives which reacted specifically with radicals (radical scavengers).

Evidence for ionic reactions in condensed systems then appeared. For example, polymerization of isobutene, which proceeds only by a cationic mechanism, was found to occur on irradiation. Also, mass spectrometric studies showed that *ion–molecule reactions* provided alternative reaction pathways.

Electron spin resonance and pulse radiolysis were then developed. These techniques enabled radicals to be observed and identified. Their concentrations could be measured and their reactions followed. Pulse radiolysis produced direct evidence of the existence of the electron, solvated in polar solvents, immediately after irradiation (maximum lifetime 600 microseconds in water). It is now realized that *free radicals, ions and excited states* may be involved in the radiation chemical reactions.

The exact physical processes are still a subject of controversy. It is known that the initial distributions of ions and radicals are not uniform. Instead, there are local regions containing different numbers of these species in varying concentrations. These are the tracks, spurs, delta rays, etc., and diffusion from them into the bulk of the material plays an important role in the reaction sequence.

Radiation chemistry is now at an exciting stage of development. The discoveries that are being made will contribute to advances in many other fields, and more industrial processes utilizing radiation will become established. The remainder of this book will help the reader to follow these advances.

LITERATURE

There is an extensive literature on radiation chemistry and in the next few pages we will give an outline of this for the reader who, after digesting our account of the fundamentals, wishes to delve further. References to particular sections of radiation chemistry are listed at the end of the appropriate chapters.

General references

Although some of the older textbooks do not give an adequate picture of the role of the electron, they do provide excellent summaries of experimental results and of the main theoretical features.

ALLEN, A. O., *The Radiation Chemistry of Water and Aqueous Solutions*, Van Nostrand, New Jersey (1961).

AUSLOOS, P. (Ed.), *Fundamental Processes in Radiation Chemistry*, Wiley, New York (1968).

CHARLESBY, A., *Atomic Radiation and Polymers*, Pergamon, Oxford (1960).

HAISSINSKY, M. (trans. TUCK, D. G.), *Nuclear Chemistry and its Applications*, Addison Wesley, Reading, Mass. (1964).

HAISSINSKY, M. (Ed.), *The Chemical and Biological Action of Radiations*, Academic, London (1961) (and several subsequent volumes in the series published by Masson, Paris).

HAISSINSKY, M. and MAGAT, M., *Selected Constants—Radiolytic Yields*, Pergamon, Oxford (1963).

HENLEY, E. J. and JOHNSON, E. R., *The Chemistry and Physics of High Energy Reactions*, Washington University Press (1969).

LEA, D. E., *Actions of Radiations on Living Cells*, Cambridge University Press (1956).

LIND, S. C., HOCHANADEL, C. J. and GHORMLEY, J. A., *Radiation Chemistry of Gases*, Chapman and Hall, London (1961).

LIVINGSTON, R. S. (Ed.), *Radiation and Photochemistry*, which is Topic 22 in *International Encyclopaedia of Physical Chemistry and Chemical Physics*, Pergamon, Oxford (1970).

SPINKS, J. W. T. and WOODS, R. J., *An Introduction to Radiation Chemistry*, Wiley, New York and London (1964).

SWALLOW, A. J., *Radiation Chemistry of Organic Compounds*, Pergamon, Oxford (1960).
VERESHCHINSKIY, I. V. and PIKAEV, A. K., *Introduction to Radiation Chemistry*, Davey, New York (1965).

Original papers

Original contributions to the field of radiation chemistry, amounting to a few thousand papers each year, are published in the general chemical literature. The specialist can find these through the normal library channels and the abstracts literature. In particular there are sections dealing with radiation chemistry in *Chemical Abstracts* and in *Nuclear Science Abstracts*.

Access to the patent literature can also be obtained through abstracts journals and patent lists, etc.

The U.S. Atomic Energy Commission sponsors many research projects on basic radiation chemical research in educational institutions and the progress reports provide interesting reading. These and other reports issued from U.S. National Laboratories can be found from *Nuclear Science Abstracts*. The Russian journal *Khimiya Vysokikh Energi*, available in English translation as *High Energy Chemistry*, and the *International Journal for Radiation Physics and Chemistry* feature radiation chemistry research. So does *Radiation Effects* which is published by Gordon and Breach (New York). The journal, *Radiation Research* and Supplements contains some papers on radiation chemistry, but is mainly devoted to radiation biology.

The U.S. Atomic Energy Commission issues translations of the *Annual Report of the Japanese Association for Radiation Research on Polymers* (now the *Osaka Laboratory* of the *Japan Atomic Energy Research Institute*). There is also the *Annual Report of the Radiation Center of Osaka Prefecture*.

Bibliographies

A bibliographic list up to December 1960 was compiled by R. W. Clarke as a Harwell report—AERE C/R 1575 and its supplements.

Most aspects of radiation chemistry are covered in ANTHONY, L. J., *Sources of Information on Atomic Energy*, Pergamon, London (1966).

Isotope Titles (Isocommerz GmbH, Lindenberger Weg 70, 1115 Berlin-Buch, German Democratic Republic) lists isotope publications (including radiation) in a large number of scientific journals and monographs and has a keyword index.

Radiation Effects in Materials and *Radioisotope Applications* are published by the Current Literature Awareness Service, Western New York Nuclear Research Center, Inc., Power Drive, Buffalo, New York.

As one of the CHEMSCAN computer-based experimental current-awareness bulletins, the Chemical Society of London is preparing lists of current titles on 'Radiation and Photochemistry'.

The Radiation Chemistry Data Centre of the University of Notre Dame maintains a comprehensive collection of data and references, and issues a weekly list of titles for specialist research workers.

Reviews

Reviews of radiation chemistry or of certain areas in this field are regularly included in *Annual Reviews of Physical Chemistry*, and from time to time in

Annual Reviews of Nuclear Science and in other review journals. *Isotopes and Radiation Technology* is a useful quarterly progress review published by the U.S. Atomic Energy Commission. There is a series edited by M. EBERT and A. HOWARD, *Current Topics in Radiation Research*, North-Holland, Amsterdam, Vol. 1 (1965) and further volumes in subsequent years, which contains reviews of predominantly biological interest, together with some which are more chemical. AUGENSTEIN, L. G., MASON, R. and QUASTLER, H. (Eds.), *Advances in Radiation Biology*, Vol. **1**, Academic Press, New York (1964) onwards often has review papers on certain aspects of radiation chemistry. There is also a quarterly, PHILLIPS, G. O. and CUNDALL, R. B. (Eds.), *Radiation Research Reviews*, Vol. **1**, Elsevier, Amsterdam (1968) onwards, and an annual—BURTON, M and MAGEE, J. L. (Eds.), *Advances in Radiation Chemistry*, Vol. I, Wiley, New York (1969) and subsequently.

The American Society for Testing and Materials list of publications should also be consulted.

Nuclear News and newsheets such as *Nucleonics News*, *Radioisotopes Report* and the *Technical News Bulletin* of U.S. National Bureau of Standards and a number of general periodicals concerning atomic energy often have paragraphs dealing with new developments in the applications of radiation.

Meetings and conferences

The meetings of a number of chemical and other societies have sessions devoted to radiation chemistry, but usually the proceedings are not published. Exceptions are the famous Geneva *Conferences on the Peaceful Uses of Atomic Energy*.

The International Atomic Energy Agency sponsors discussion meetings on various topics connected with radiation chemistry, and their list of publications should be referred to. Particularly interesting is *Radiation Chemistry and its Applications*, Technical Reports Series No. 84, International Atomic Energy Agency, Vienna (1968). There are also the 1959 Warsaw conference on *Large Radiation Sources in Industry* (two volumes), the 1963 Salzburg conference on *Industrial Uses of Large Radiation Sources* (two volumes) and the 1969 Munich symposium on large radiation sources and accelerators in industrial processing.

Abstracts of papers presented at meetings of the Radiation Research Society are published in *Radiation Research*. Abstracts from meetings of the American Nuclear Society are published in *Transactions of the American Nuclear Society*.

Proceedings of the International Congress of Radiation Research 1958, Burlington, Vermont, is published as Supplement **1** (1959) of *Radiation Res.*; the 1962 Harrogate Congress on Radiation Research as EBERT, M. and HOWARD, A. (Eds.), *Radiation Effects in Physics, Chemistry and Biology*, North-Holland, Amsterdam (1963); the 1966 Cortina Third International Congress of Radiation Research as SILINI, G. (Ed.), *Radiation Res.*, North-Holland, Amsterdam (1967) and the Fourth International Congress on Radiation Research at Evian, France, in 1970 will no doubt be published.

Published proceedings of specialist conferences on radiation chemistry include:

Radiation Chemistry, *Discussions of the Faraday Society* No. **12**, the Leeds meeting (1952).

All-Union Conference on Radiation Chemistry (Moscow, 1957), Consultants Bur. Inc. N.Y. (1959).

BURTON, M., KIRBY-SMITH, J. S. and MAGEE, J. L. (Eds.), *Comparative Effects of Radiation*, Wiley, N.Y. (1960)—the meeting in San Juan, Puerto Rico in Feb. 1960.

The Tihany Symposium on Radiation Chemistry (1962) and the second conference and volume (1966), Akademiai Kiado, Budapest.

Proceedings of the Nuclear and Radiation Chemistry Symposium, 1964 (Bombay) and 1967 (Poona) published by the Chemical Council, Department of Atomic Energy, Government of India.

Discussions of the Faraday Society No. 36, *Fundamental Processes in Radiation Chemistry* (1963)—the Notre Dame conference.

HART, E. J. (Ed.), Basic Mechanisms in the Radiation Chemistry of Aqueous Media, *Radiation Research* Supplement 4 (1964)—a conference held at Gatlinburg, U.S.A., in 1963.

PHILLIPS, G. O. (Ed.), *Energy Transfer in Radiation Processes*, Elsevier, Amsterdam (1966)—an International Symposium on chemical, physical and biological aspects held at Cardiff in January 1965.

EBERT, M., KEENE, J. P., SWALLOW, A. J. and BAXENDALE, J. H. (Eds.), *Pulse Radiolysis*, Academic, London (1965)—a conference held in Manchester in 1965.

HART, E. J. (Ed.), *Solvated Electron*, Advances in Chemistry Series No. **50** by the American Chemical Society, a Symposium sponsored by the American Chemical Society and held at Atlantic City in September, 1965.

HART, E. J. (Ed.), *Radiation Chemistry*. The International Conference at Argonne in August, 1968, is covered by two further volumes, Nos. 81 and 82 in Advances in Chemistry Series, American Chemical Society. R. F. GOULD is General Editor of the Series.

JOHNSON, G. R. A. and SCHOLES, G. (Eds.), *The Chemistry of Ionisation and Excitation*, Taylor and Francis, London (1967)—the Newcastle meeting.

STEIN, G. (Ed.), *Radiation Chemistry of Aqueous Systems*, Wiley, London (1968), is the proceedings of the Farkas Symposium held in Jerusalem, December 1967.

Papers presented at the Symposium on Photochemistry and Radiation Chemistry held at Natick in April 1968 were published in a special issue of the *Journal of Physical Chemistry* (October, 1968).

Japan Conferences on Radioisotopes are organized periodically by the Japan Atomic Industrial Forum.

The Gordon Conferences in the United States and Miller Conferences in Europe are arranged for a limited number of selected research workers. In order to encourage free and uninhibited discussion the proceedings are not published. The Radiation Research Laboratory at Notre Dame has also held informal conferences on the radiation chemistry of water and the contributions and discussion have been published in report form through the United States Atomic Energy Commission.

Films

A useful adjunct to any introductory course in Radiation Chemistry is the short (13-min) colour film *Radiation Effects in Chemistry*, produced in the United States in 1964. It indicates the basic role of the initial ionization and excitation in the subsequent chemical reactions. Some of the sophisticated techniques, such as pulse radiolysis and electron spin resonance with *in-situ* beam-irradiation are shown in operation.

Chapter 2

Interaction of Radiation with Matter

TYPES OF RADIATION

The different types of high-energy ionizing radiation may be classified according to the distance they penetrate through material. High-energy electromagnetic radiations, such as 1 MeV (0.16 pJ) γ-rays, and the

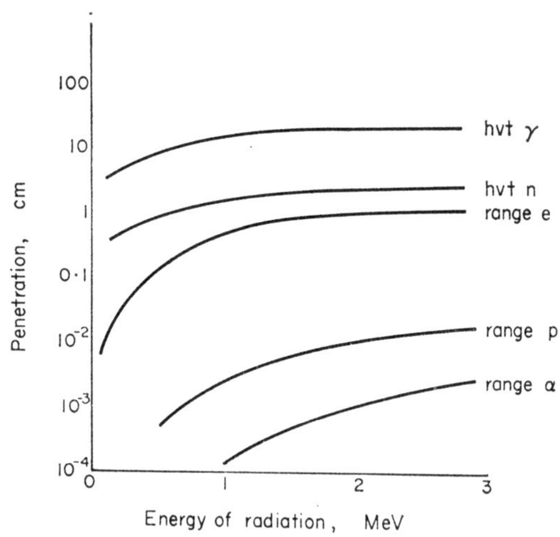

Fig. 2.1 Half value thickness (hvt) or range of various radiations in material of unit density. The hvt is the thickness of the medium which causes a reduction of the radiation intensity to one-half.

neutral particles—neutrons—may penetrate through about a metre of solid or liquid. On the other hand, charged particles—electron beams, β-particles, proton beams and α-particles—penetrate no more than a few millimetres in solids or liquids, or some centimetres in gases (see Fig. 2.1 and Table 2.1). X-rays from the usual x-ray sets (20–50 kV), or

γ-rays of similar energy, penetrate intermediate distances. Except for neutrons, radiation interacts with orbital electrons. Therefore, deposition of energy is dependent on electron density, for which mass density is a good approximation.

Much of the knowledge of the passage of radiation through materials has been obtained from experiments using cloud or bubble chambers, or photographic emulsions. Ionization tracks and collisions can be observed by these means. However, more radiation chemistry investigations, both experimental and theoretical, have been made on water than on any other material. Therefore we will draw many of our examples and comparisons from the aqueous field. Other materials will differ in detail, but the broad principles will be the same.

We will see that the important features are (1) the amount of energy deposited by the radiation and (2) its microscopic spatial distribution, in the material being irradiated. The penetration determines the location within the material of this deposited energy and of the resultant ionization.

Table 2.1. Penetration of 1 MeV radiations.

Type of Radiation		Penetration In water cm	Penetration In air cm
X or gamma	hvt	10	7000
Beta or electron beam	Range	0.5	400
Alpha	Range	0.0005	0.2
Proton	Range	0.002	2.3
Neutron	hvt	1.5	Large
Fission fragment (100 MeV)	Range	0.01	2.5

hvt = half value thickness

Electromagnetic radiation

X-rays and γ-rays are electromagnetic radiation similar to visible or ultraviolet light, but of much shorter wavelength, and hence greater photon quantum energy.

X-RAYS *X-rays* are produced when a target is hit by a stream of fast-moving electrons, such as those which have been accelerated by an electric field in a vacuum. The higher the atomic number of the target, the more efficient the production of x-rays. The x-radiation has an energy continuum from zero up to a maximum, which is identical with that of the electron beam. Superimposed on this continuum are several peaks, of wavelengths characteristic of the target material (Fig. 2.2). Sometimes a plate of metal is interposed to filter out the longer-wavelength (lower-energy) radiation and improve the 'quality' of the radiation.

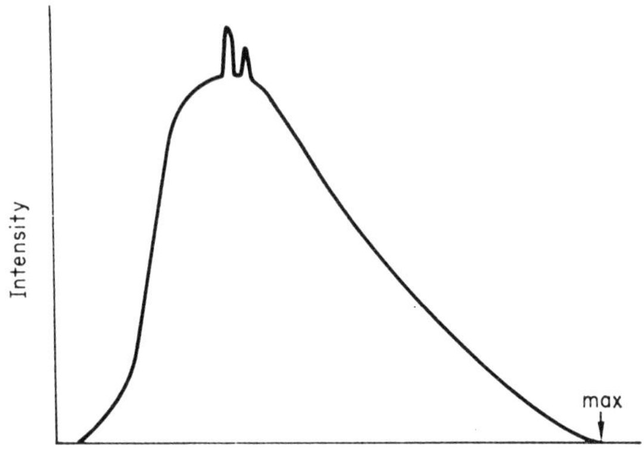

Fig. 2.2 Energy spectrum for x-rays. The peaks are characteristic of the target.

GAMMA-RAYS *Gamma-rays* are emitted from many radioactive nuclei. Unstable atomic nuclei produced during nuclear transmutations often emit excess energy as γ-rays. They have discrete energies, although the γ-emissions from some radioactive nuclei are quite complex.

ABSORPTION High-energy electromagnetic radiations lose a certain fraction of their energy in passing through each unit thickness of materials, and thus they never have a definite range. At any depth of penetration there is still some radiation present, although the intensity becomes vanishingly small at great depths. The absorption depends on the energy

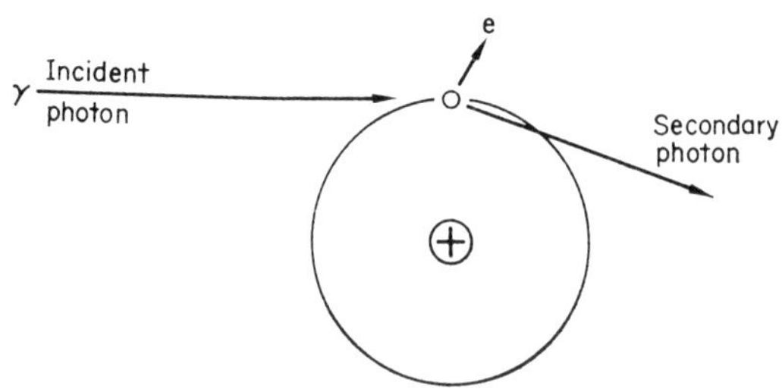

Fig. 2.3 Compton scattering of γ-photon—10 MeV to 10 keV.

of the γ-photon, on the density of the material and on the atomic numbers of its components. After passing through 10 cm of water 1 MeV γ-radiation has lost about 50% of its initial intensity.

Initially, the main process by which 1-MeV γ-photons lose energy is *Compton scattering*. In this process, part of the gamma photon energy is used to eject an orbital electron with considerable kinetic energy, and the remainder of the energy continues in the scattered photon (Fig. 2.3).

For photon energies below about 60 keV (9.6 fJ) the *photoelectric* effect

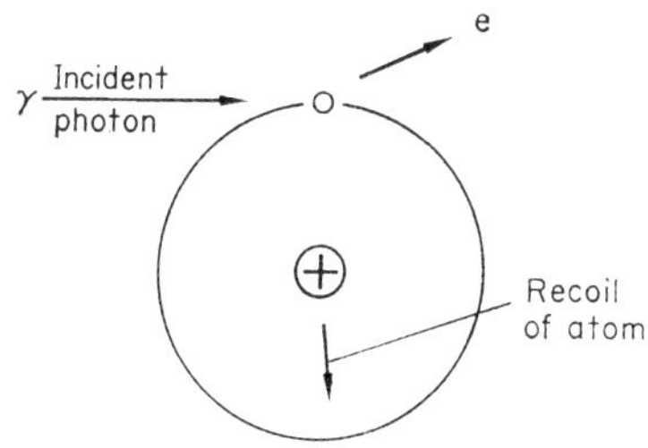

Fig. 2.4 Photoelectric effect—below 60 keV.

becomes the most important process. All the energy of the photon is used in ejecting an electron, usually from an inner orbital of the atom (Fig. 2.4). The filling of an inner orbital vacancy from an outer orbital provides further energy, which often appears as a number of electrons from the one atom—the *Auger* effect—or as characteristic fluorescent x-radiation.

Another process—production of an electron–positron *pair*—is un-

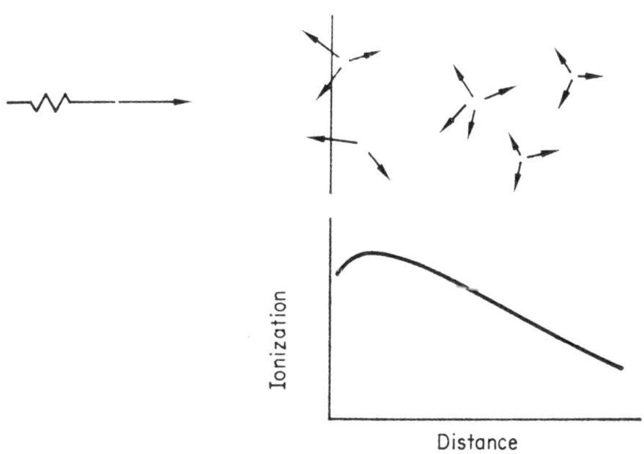

Fig. 2.5 Ionization caused by γ-irradiation as a function of depth of penetration in material of unit density.

important for photons of energy less than 10 MeV (1.6 pJ), except for heavy elements. Photon, nuclear and other reactions involving atomic nuclei are also unimportant except at very high energies. *Rayleigh*, or coherent, scattering is important only at energies less than about 50 keV (8 fJ) and, in any case, involves little deposition of energy.

Near the surface of the irradiated material there is no scatter from previous layers. Therefore the effects will be somewhat different from those in the bulk of the material. This phenomenon is called 'build up'. It is not until about 1 cm of water has been penetrated by 1 MeV γ-radiation that conditions of spatial equilibrium are reached (Fig. 2.5).

Light charged particles

Electron irradiation

Beta-particles and electron beams are high-energy (high-velocity) electrons.

BETA-PARTICLES *Beta-particles* arise from the decay of many radioactive nuclei. There is a maximum energy characteristic of the radioisotope, but most of the radiation is of lower energy (Fig. 2.6). The mean energy is 0.4 times the maximum energy.

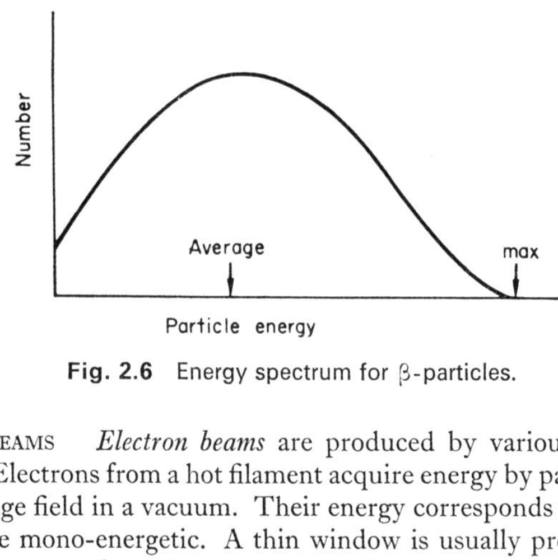

Fig. 2.6 Energy spectrum for β-particles.

ELECTRON BEAMS *Electron beams* are produced by various accelerating machines. Electrons from a hot filament acquire energy by passing through a high-voltage field in a vacuum. Their energy corresponds to this voltage and they are mono-energetic. A thin window is usually provided so that the beam may pass from the vacuum chamber and into the system being irradiated.

We have already seen that high-energy electrons are produced by Compton and photoelectric processes in material irradiated by γ-rays.

ABSORPTION The range of high-energy electrons in a material is directly proportional to their energy and inversely to the density of the material. For 1-MeV electrons the maximum penetration in water is 5–6 mm (Fig. 2.7). As with γ-radiation, electrons do not produce their maximum effect at the surface of the irradiated material.

The incident electrons eject orbital electrons by the process of *Coulombic* interaction, producing ionization. Thus the primary electrons are slowed down and cascades of secondary, tertiary, etc., electrons are produced, as shown in Fig. 2.8.

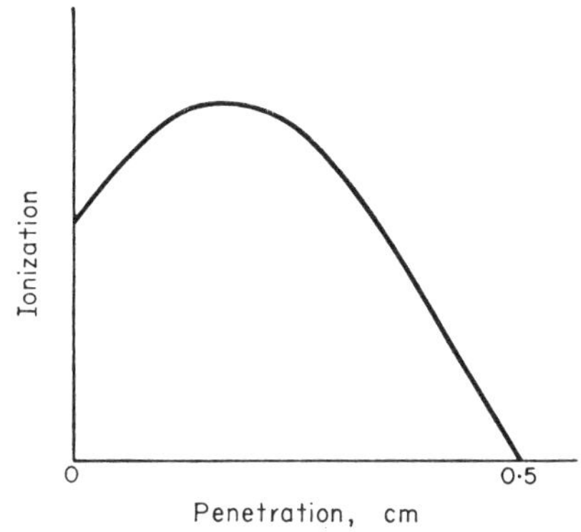

Fig. 2.7 Ionization produced by 1 MeV (0.16pJ) electrons as a function of depth of penetration.

The ionization density is higher than for γ-rays, but still not very high, except towards the ends of the tracks, where the electrons are moving more slowly and make more frequent collisions.

More rarely, the high-energy electrons interact with nuclei, giving x-

Fig. 2.8 Part of cascade produced by a single electron.

rays called *Bremstrahlung*, which may be later absorbed by the material. A small amount of energy is also lost as *Cerenkov* radiation—the spectacular blue glow produced as electrons slow down to the speed of light in material of refractive index greater than one.

Heavy charged particles

Alpha-particles, proton beams, etc., consist of the positively charged nuclei of atoms shorn of their orbital electrons. They have short ranges, and hence produce very intense ionization along their tracks.

ALPHA-PARTICLES *Alpha-particles* are produced from some radioactive heavy elements. They consist of helium nuclei, ${}^4_2He^{2+}$, and have discrete energies.

PROTONS AND DEUTERONS *Protons* (${}^1_1H^+$) and *deuterons* (${}^2_1D^+$), are the nuclei of ordinary and heavy hydrogen atoms respectively. When produced directly by accelerating machines they are mono-energetic, but when produced by the slowing down of fast neutrons in hydrogenous material such as water or paraffin wax they have a range of energies.

FISSION FRAGMENTS *Fission fragments* are formed whenever fissile materials, such as uranium-235 or plutonium-239, capture neutrons, for example in a nuclear reactor. Fission fragments are produced when the compound nuclei split in two. They are the nuclei of elements in the middle of the periodic table, partially shorn of electrons so that initially they each carry a net charge of about +20. They have enormous energies (up to about 100 MeV = 16 pJ) and form extremely dense ionization tracks.

ABSORPTION The range of heavy particles in water is very short. For 1-MeV particles the ranges in water are 5 microns (5 μm) for alpha particles and 30 microns (30 μm) for protons. The range of fission fragments is about 0.1 mm, i.e. 100 μm.

The processes by which heavy charged particles lose energy are similar to those for light charged particles, leading to ionization of the molecules of the medium. Since all the energy is deposited in a short distance, the ionization density is very high, particularly towards the end of the range.

Neutrons

Neutrons arise from fission events in atomic reactors or when α-particles, accelerated charged particles, high-energy x-rays or γ-rays impinge on certain targets.

ABSORPTION Because they are uncharged, neutrons do not interact with orbital electrons. They lose their energy by elastic collision with atomic nuclei—the lighter the nuclei, the greater the amount of energy transferred. Fast neutrons penetrate several centimetres in water, and along their tracks produce knock-on protons, which act in just the same way as would accelerated protons in the same locations.

After several collisions the neutrons have been slowed down and no longer displace atoms. Finally they reach thermal energies (0.025 eV = 0.004 aJ) and may be captured by certain elements. Following the capture, excess energy may be emitted as a γ-ray and the capturing atom may recoil out of its molecule. This recoiling atom is called a 'hot atom' and acts like a heavy charged particle. Certain elements, after capturing a neutron, emit α-particles. These elements are sometimes called 'energy converters'. Thus boron-10 after absorbing a neutron decomposes into lithium-7 and an α-particle.

$$^{10}_{5}\mathrm{B} + {}^{1}_{0}\mathrm{n} \longrightarrow {}^{7}_{3}\mathrm{Li} + {}^{4}_{2}\alpha \qquad (2.1)$$

Reactor radiation

Reactor, or *pile, radiation* is a mixture of γ- and neutron radiation. This makes the chemical effects of the radiation rather difficult to interpret. hence it is not suitable for basic studies.

Induced radioactivity

Many elements in a neutron flux, particularly a slow neutron flux, become radioactive. The nuclei of the elements absorb the neutrons, and the resultant compound nuclei are unstable and emit radiation. For example:

$$^{23}_{11}\mathrm{Na} + {}^{1}_{0}\mathrm{n} \longrightarrow {}^{24}_{11}\mathrm{Na} \qquad (2.2)$$

$$^{24}_{11}\mathrm{Na} \xrightarrow[\text{15 hr}]{\text{half life}} {}^{24}_{12}\mathrm{Mg} + {}^{0}_{-1}\beta + \gamma \qquad (2.3)$$

This radiation may cause radiation chemical effects in the material. In any case it is a potential health hazard. In certain circumstances other radiation can induce radioactivity in materials. Photon and electron radiation of energy less than 1.66 MeV cannot induce any radioactivity whatsoever in any material. Beyond 2 MeV they can induce radioactivity in some cases.

PHYSICAL ASPECTS

Ionization and excitation

We have, up to this point, considered *ionization* as the phenomenon which results when high-energy radiation interacts with matter. However, during a glancing collision insufficient energy may be transferred to eject an electron completely, i.e. to effect ionization. If an electron is raised to a higher orbital the molecule will be electronically *excited*. The secondary electrons towards the end of their paths are more likely to cause excitation than ionization. Consequently there is a zone of excitation surrounding the zone of ionization along the track of an ionizing particle.

If one electron is ejected and another in the same atom is excited an *excited ion* results. If two partners of an ion pair recombine, i.e. an electron and its parent positive ion, an excited molecule will be produced. We will see that a large proportion of the energy deposited in a material appears as heat, i.e. as other forms of excitation (vibrational, rotational and kinetic), but electronic excitation is probably more important because it can lead to chemical reaction. When we talk about excited states we will usually mean electronically excited, and indicate this by an asterisk, e.g. M*.

Photon and light particle irradiations

Let us consider what happens when 1 MeV γ-radiation passes through a substance. The electrons from the Compton scattering of 1 MeV γ-photons have an average energy of 400 keV. This means that the electrons produced by Compton scattering and by the photoelectric effect—often called *delta-rays*—will act just like β-particles in bringing about further ionization and excitation.

These ionization processes all produce electrons and leave behind positively charged ions of the target material,

$$M \rightsquigarrow M^{\cdot +} + e^- \tag{2.4}$$

and excited molecules.

$$M \rightsquigarrow M^* \tag{2.5}$$

(The fact that a reaction has been brought about by radiation is often indicated by a wavy arrow.)

The electrons finally reach energies below the lowest ionization or electronic excitation level of the major components of the substance. These *sub-excitation electrons* lose energy much more slowly. This means that they can possibly affect some minor components and impurities with low-energy excitation levels. In general they slow down finally to thermal energies.

$$e^- \longrightarrow e^-_{\text{thermal}} \tag{2.6}$$

It will have been observed that the electrons produced in a substance by γ-radiation undergo the same sorts of interactions as do the electrons from β-particles or electron beams. Essentially, then, γ-rays, x-rays, beta particles and electron beams give the same sort of track microscopically. They do differ in penetration, but at the location of the interaction with the components there is little difference between them.

Time scale

Although a number of doubtful points remain, the main features of the interaction of radiation with liquid water are clear. The time scale of these events is given in Table 2.2.

Table 2.2. Time scale of events in liquid water following passage of 1 MeV electron or gamma radiation.

Seconds	Event
10^{-18}	Particle traverses water molecule.
10^{-15}	Time between successive ionizations.
10^{-13}	Molecular vibration and dissociation.
5×10^{-13}	Electron moderated to thermal energies.
10^{-12}	Radical, etc., moved one jump in diffusion.
10^{-11}	Electron solvated.
10^{-8}	Allowed excited transition. Intra-site reactions completed.
10^{-7}	Fastest reactions with mM solute.
10^{-4}	Radical–radical reactions.
10^{-3}	Phosphorescence from triplet states.
$> 10^{-3}$	Most early reactions are complete, a few slow reactions are proceeding, some post-irradiation reactions continue for hours or weeks.

Physical picture

The widely separated events along the *main track*, or δ-ray *branch tracks* formed by energetic secondary or tertiary electrons, are called *isolated spurs*. For materials composed of light atoms, such as water, most isolated spurs contain 2–4 species and the energy involved is only about 10–25 eV (1.6–4 aJ). We will use the term *isolated spurs* to avoid confusion with the word *spur*, which is often used for other events as well.

If the δ-ray electron has energy less than about 5 keV (800 aJ), then the spurs will be formed close together and will overlap. These *short tracks* will be cylinders quite different from the isolated spurs of the main and branch tracks.

Towards the ends of the tracks, that is as the electron energies fall to between about 500 and 100 eV (80 and 16 aJ), there will be even denser pear-shaped regions of ionization which look like large spurs, containing 100–200 eV (16–32 aJ) of energy (mostly) and a relatively large number of species. These have been called *blobs*. This is illustrated schematically in Fig. 2.9.

From each 1-MeV electron entering a material there are something like 10^5 species formed. The energy is apportioned in water in the ratio isolated spur: short track: blob as 67%:22%:11%.

The important thing about this picture is that there is a *non-uniform* distribution of species. Those formed in close proximity, as in blobs or short tracks, are better able to react together rather than with other substances present.

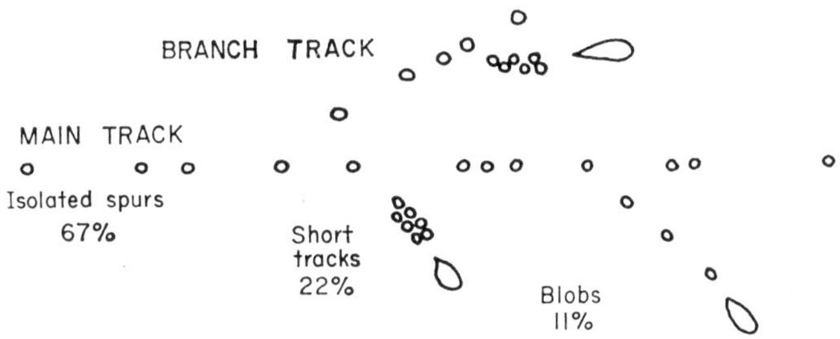

Fig. 2.9 Schematic diagram of distribution of events (ions and excited molecules) in water following passage of 1-MeV electron.

Heavy charged particles

On the other hand, for heavy charged particles, whether they be accelerated protons, recoil ions or α-particles, the ionization density is so much higher that the events can be considered as producing a series of overlapping stubby short tracks and blobs (Fig. 2.10). Also, there will be

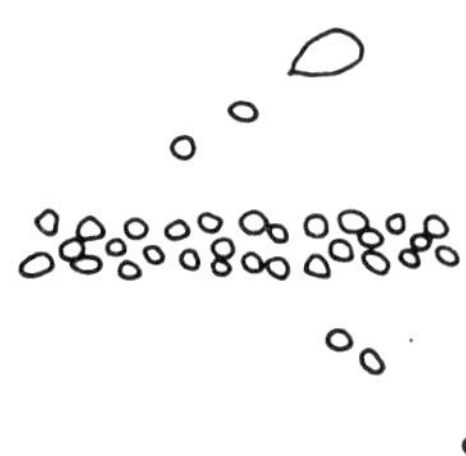

Fig. 2.10 Schematic diagram of distribution of events (ions and excited molecules) in water following passage of 1-MeV α-particle.

a smaller proportion of isolated spurs produced by branch-track electrons. This in turn leads to rather different chemical consequences.

GASES In gases the electrons formed are able to escape more readily from their parent positive ions, so there is much less likelihood of their recombining. This means that, in gases, there are not the differences between electromagnetic and light particle irradiation, on the one hand, and heavy particle irradiation, on the other, that are found in condensed phases.

SOLIDS In solids, or gel materials, such as some biological systems, the proximity of events can be very important. Thus a minimum finite number of ionizations or excitations in a particular region might be necessary for the death of an organism. In such cases some phenomena, such as the Auger effect, Rayleigh scattering and elastic scattering, may become important.

Linear Energy Transfer (L.E.T.)

An important concept is that of *Linear Energy Transfer*—usually abbreviated to L.E.T. This is the energy deposited per unit length of

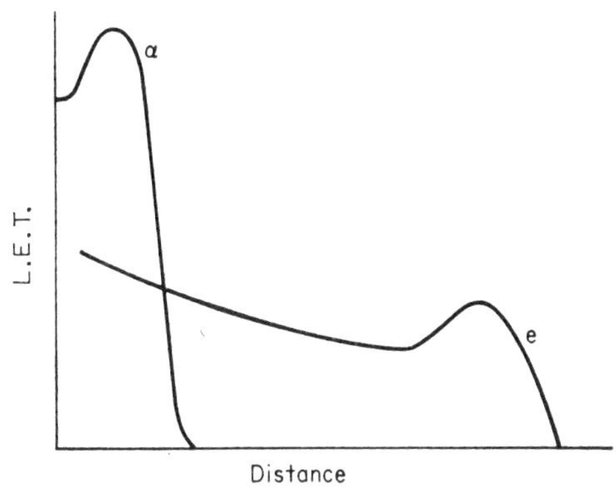

Fig. 2.11 L.E.T. (Linear Energy Transfer) as a function of distance for α-particle and high-energy electron.

track; it is usually expressed as eV per Å, eV per mμ, or aJ per nm (1 eV per Å = 1.6 aJ per nm), and is different for each type of radiation. L.E.T. as a function of distance is shown in Fig. 2.11 for an α-particle and a high-energy electron.

Table 2.3. Initial L.E.T. values for various types of radiation.

Radiation	Initial L.E.T. eV/Å	aJ/nm
Co-γ, 1–2 MV electrons	0.02	0.032
250 kV x-rays	0.10	0.16
10 kV x-rays	0.20	0.32
2 MeV protons	1.7	2.7
delta	0.8–3.0	1.3–4.8
Po alpha	8.8	14.1
Li recoil	10–17	16–27
Fission fragment	700	1120

From our discussion, it can be seen that, for electron and γ-radiation in water, there are three regions, each with a characteristic ionization density or L.E.T. value, hence an average L.E.T. value has little meaning. The initial L.E.T. of the radiation in the material is rather more significant. Some typical values are given in Table 2.3.

Radiation units

Throughout the foregoing we have laid stress on *energy deposited* by the radiation, i.e. the energy absorbed by the system being studied. This is because the amount of *absorbed energy*, in conjunction with the nature of the material being irradiated, determines the number of molecules which are ionized or excited.

For a given material the number of these events is proportional to the energy absorbed, and does not seem to depend very much upon the nature or energy of the radiation used, at least for energies greater than 20 keV (3.2 fJ).

In radiation chemistry the *absorbed dose* (usually given the symbol D) is most often expressed in terms of a unit called the *rad*, corresponding to 100 ergs (10 μJ) of energy absorbed per gram of material irradiated. In Chapter 3 we show how the absorbed dose is measured. Another unit often used is *electron volts per millilitre* (1 eV/ml = 0.16 aJ/cm^3) or *electron volts per gram* (1 eV/g = 0.16 aJ/g). An allowable abbreviation for rad is rd, but this is seldom used.

In gases it is possible to measure the ion pair yield directly. The *W value*—the average energy expended in a gas per ion pair formed—varies from about 20 to 45 and is 33.7 for dry air (eV).

Some W values are given in Table 5.1. The W value for a mixture of gases must be determined experimentally and cannot be calculated from the composition of the mixture. W values are usually expressed as electron volts (1 eV = 0.16 aJ).

For x-rays or γ-rays the ionization, or *exposure*, at any point is expressed as *roentgen* (R). One roentgen produces a total of 2.58×10^{-4} coulomb of electrical charges on all the ions of one sign in one kilogramme of air. This is identical with the previous definition of one e.s.u. of charge per 0.001293 g of air (1 cm^3 at 0°C and 760 mmHg pressure). An exposure of one roentgen corresponds to an absorbed dose in air of 86.9 erg/g (8.69 μJ/g).

One roentgen of 1 MeV γ-radiation deposits 96.5 erg (9.65 μJ)/g in water, but this figure varies with the energy of the radiation and with the material being irradiated. The rad is the measure of energy absorbed in the system and is preferable to the roentgen. The roentgen is rarely used in radiation chemistry. The absorbed dose in a system in rad is numerically the same as the exposure in roentgen, within a few per cent. It will be realized that they are not dimensionally the same.

There are some biological systems which are sensitive to doses of the order of a few rads. In radiation chemical studies solutions are often given doses of a few kilorads (krad) while organic liquids and solids might be given a few megarads (Mrad).

CHEMICAL ASPECTS

Essentially all radiations give the same number of species (ion pairs, etc.) in a given system for a given dose. The spatial distribution and concentration of these in short tracks and blobs is such as to favour *intra-track* reactions in condensed phases. Reactions between species within the isolated spurs are much less probable. The species which do not react diffuse away from their birthplace and can then react with species from another site, or with a component of the material being irradiated.

Free radicals

The electron, its adducts with other stable substances, the parent positive ion and radical pairs formed by homolysis of excited molecules are all *free radicals*. Their common characteristic is that they all have an unpaired electron, and usually they are very reactive. Many of the reactions of radiation chemistry are free radical reactions.

To draw attention to the fact that a species is a free radical we will indicate the presence of the unpaired electron by a dot on the chemical formula, e.g. R·.

Some free radical reactions are extremely fast and have high rate constants, of the order of 10^{10}–10^{11} M^{-1} s^{-1}† Radical–radical reactions in particular often occur at every encounter, i.e. are 'diffusion controlled'.

When a radical reacts with an ordinary stable molecule or ion the immediate product is a radical.

$$R\cdot + M \longrightarrow RM\cdot \quad \text{or} \quad RX + M'\cdot \qquad (2.7)$$

This radical may also be reactive, and the only way in which the series can be terminated is by reaction between two radicals,

$$A\cdot + A\cdot \longrightarrow A_2 \qquad (2.8)$$

or by reaction of a radical with an element of variable valence, such as iron in the form of ferrous or ferric ions.

$$A\cdot + Fe^{2+} \longrightarrow A^- + Fe^{3+} \qquad (2.9)$$

† Second order rate constants will be expressed in units of litre $mole^{-1}$ $second^{-1} \equiv$ Molar $concentration^{-1}$ $s^{-1} \equiv dm^3\ mol^{-1}\ s^{-1}$.
$1 M^{-1} s^{-1} = 10^3\ cm^3\ mol^{-1}\ s^{-1} = 10^{-3}\ m^3\ mol^{-1}\ s^{-1} \approx 1\ kg\ mol^{-1}\ s^{-1}$ in dilute aqueous solution.

Sometimes the adduct of a primary radical with a stable component can exist for a finite time as an intermediate species. In fact, certain free radicals are stable for years. In other cases a true transition state, or activated complex, is formed, and this decomposes immediately.

Intra-site reactions

The electrons, being more mobile, are better able to diffuse away from their birthplace than are the heavier positive ions. One school of thought (Samuel–Magee) considered that, in the case of irradiated water, these electrons did not escape from the electrostatic attraction of their parent, positive ions. Another school (Lea–Gray–Platzman) thought that most of the electrons escaped altogether.

It is now known that the proportion that are attracted back, each to its parent positive ion, and undergo *geminate recombination* depends inversely on the dielectric constant of the material. The proportion is low in water and polar materials and high in organic liquids with low dielectric constants. Other factors, such as the relaxation times of the two ions and their respective environments, their diffusivity and the efficiency of electron trapping, may modify the simple dependence on static dielectric constant. The high frequency dielectric constant would be more appropriate for small distances of separation of the ions and perhaps for unrelaxed ions. The process of geminate recombination is not completely understood.

The electrons that are attracted back are neutralized, and this gives the original molecules in an excited state.

$$M^{\cdot +} + e^- \longrightarrow M^* \qquad (2.10)$$

These excited molecules may revert to the ground state by radiationless transition or by luminescence—emitting a quantum of energy—giving no net reaction.

$$M^* \longrightarrow M + h\nu \qquad (2.11)$$

Alternatively, they may undergo homolysis to give two free radicals,

$$M^* \longrightarrow A\cdot + B\cdot \qquad (2.12)$$

or heterolytic scission to give an ion pair,

$$M^* \longrightarrow A^+ + B^- \qquad (2.13)$$

or molecular elimination.

$$M^* \longrightarrow C + D \qquad (2.14)$$

Frequently D is molecular hydrogen (H_2). It can be seen that an excited molecule may undergo these and other reactions which are characteristic of its excited state and which differ from those of the ground state molecule.

The parent positive ion may dissociate before an electron can return, giving a more stable ion and a free radical,

$$M^{\cdot +} \longrightarrow A^{+} + B^{\cdot} \qquad (2.15)$$

or undergo an ion–molecule reaction.

$$M^{\cdot +} + M \longrightarrow MD^{+} + E^{\cdot} \qquad (2.16)$$

$$M^{\cdot +} + M \longrightarrow M_2^{\cdot +} \qquad (2.17)$$

These various species may further react together, re-forming the original material,

$$A^{\cdot} + B^{\cdot} \longrightarrow M \qquad (2.18)$$

or new molecular species by disproportionation (often hydrogen atom transfer)

$$A^{\cdot} + B^{\cdot} \longrightarrow C + D$$

or combination,

$$2A^{\cdot} \longrightarrow A_2 \qquad (2.19)$$

$$A^{\cdot} + M \longrightarrow MA^{\cdot} \text{ (or other products)} \qquad (2.20)$$

or hydrogen abstraction.

$$A^{\cdot} + M \longrightarrow AH + R^{\cdot} \qquad (2.21)$$

The returning electron may react with some of the species,

$$A^{\cdot} + e^{-} \longrightarrow A^{-} \qquad (2.22)$$

$$A^{+} + e^{-} \longrightarrow A^{\cdot} \qquad (2.23)$$

$$M + e^{-} \longrightarrow M^{\cdot -} \qquad (2.24)$$

and these may neutralize the parent positive ion.

$$A^{-} + M^{\cdot +} \longrightarrow MA^{\cdot} \qquad (2.25)$$

$$M^{\cdot -} + M^{\cdot +} \longrightarrow 2M \text{ or } M_2 \qquad (2.26)$$

There will be examples of these and similar reaction types in later chapters.

There will be more intra-site reactions in blobs and short tracks, i.e. in regions of high L.E.T., than there will be in isolated spurs. For heavy particle irradiations, blobs and short tracks predominate.

Reactions of the diffusing species

The species which escape from the spurs, blobs and short tracks are called *primary species*. They are not necessarily the earliest species formed in the system, but are the earliest which are readily accessible chemically.

The more stable species will diffuse away and undergo reactions similar to the above with other components of the system.

$$A\cdot + S \longrightarrow AS\cdot \qquad (2.27)$$

The electron may also diffuse away while being thermalized. It may then fall into a physical trap (lattice defect, impurity centre, etc.). In a polar solvent, such as water, alcohol, liquid ammonia or an amine, the electrostatic field of the electron may orient the solvent dipoles to form a chemical trap. These solvated electrons are very important in such solvents.

$$e^- + \text{solvent} \longrightarrow e^-_{\text{solv}} \qquad (2.28)$$

A thermal electron or solvated electron may react with an electron scavenger.

$$S + e^- \longrightarrow S\cdot^- \qquad (2.29)$$

The resulting negative ion may be fairly stable, or may disintegrate immediately (dissociative electron capture).

It is thought that sub-excitation electrons and electrons which have just reached thermal energies might be capable of undergoing chemical reactions. These electrons have been called *dry electrons* to distinguish them from solvated electrons. Because they exist for less than 10^{-11} seconds it is difficult to establish their properties but some phenomena can be explained by assuming they have a higher diffusivity and chemical properties somewhat different from those of solvated electrons.

Cage effect

Many of the reactions which we have been describing, especially those involving free radicals, are highly exothermic and produce excess energy. In a condensed phase the reactants are closely surrounded by other molecules, and these are frequently able to take away this excess energy. Therefore, when dissociative reactions occur, the fragments are able to recombine, hence there is less permanent dissociation than in the gas phase. In polar materials ions may be stabilized by solvation, and this can lead to further differences.

Final products

The products of the reactions which we have described may undergo further reactions. Finally, somewhat less than a second after the radiation has passed through the system all the very unstable species will have decomposed and all the very reactive species will have reacted. The system will have established chemical equilibrium, and the stable products can then be identified and measured.

There is often a great deal of recombination, so that the net decomposition is only a portion of the gross initial decomposition.

Post-irradiation products

There may remain some species that continue to decompose or react for hours or months after the irradiation ceases. They will give rise to post-irradiation products. Peroxides, which are often found in irradiated systems containing oxygen, sometimes decompose slowly. Trapped radicals in solids react at higher temperatures. In aerated, aqueous solutions hydrogen peroxide is formed and may react slowly, e.g. with ferrous ion.

$$H_2O_2 + Fe^{2+} \longrightarrow Fe^{3+} + \cdot OH + OH^- \qquad (2.30)$$

Radiation chemical yield or G value

The amount of a product, expressed as molecules formed for each 100 eV (16 aJ) of energy absorbed by the system, is termed the *G value* and is almost universally used to express radiation chemical yield. Expressed in another way, there is a yield of G moles for 9.65 MJ of energy absorbed, or for 965 kg Mrad. We will use the convention G (product). Where we wish to indicate the quantity of irradiated material which has reacted, per 100 eV, we will include a minus sign, i.e. G (−material). To indicate a yield of a 'primary species', usually a radical species, we will write the species as a subscript, thus G_H, although some authors prefer to use the lower case symbol g (H).

It is apparent that one must distinguish between the radiation chemical yield or G value of free ions, which are available in the bulk of the material, and the yield of geminate ions, which are available in the sites. Similar considerations apply to the radical and excited species.

Most G values lie between 0.1 and 10. Higher values usually indicate a chain reaction. About 20 eV (3.2 aJ) of radiation is absorbed for each chemical bond broken. Most chemical bond strengths are about 2 eV (0.32 aJ). The remainder ($\approx$18 eV) appears as heat.

In the Appendix there is a typical example of how the G value is calculated from a product yield.

Scavengers

Sometimes *scavengers* are deliberately added to a system to react with primary radical species. Some scavengers are specific for one particular species such as the electron and, if a stable product is formed in quantitative yield, it can be used to measure the G value of that reactive species. Other scavengers are less specific, and interpretation of product yields in terms of the primary species can be quite complex.

Scavenger concentrations of more than about 10^{-4} molar will react with all the diffusing species produced in the radiolysis of aqueous solutions. If high concentrations (over about 0.1 M) of scavengers are used it is possible to scavenge the primary species in the sites, thus inhibiting intra-site reactions. At concentrations lower than 10^{-4} M scavengers will react with only some of the diffusing species, allowing the remainder to react with each other. The effect of varying scavenger concentration is shown in Fig. 2.12.

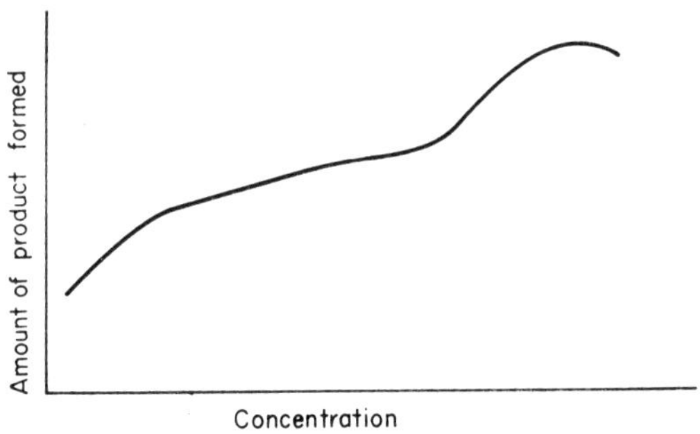

Fig. 2.12 Amount of product formed from scavenger as a function of scavenger concentration.

Sometimes scavengers are added to a system to remove certain primary species, preventing some reactions and allowing others to proceed.

Primary and secondary products

The final, stable, end products, which result from reaction between primary species and substrate components or scavengers present in the system before irradiation, are called *primary products.* As the irradiation proceeds the primary products build up and may reach such a concentration that they compete with the original components for primary species. The resultants of the reaction of primary products with primary species are called *secondary products.*

Experimentally, primary and secondary products may readily be distinguished by plotting the amount of product against dose. The line for primary products will pass through the origin, whereas that for secondary products will intersect the dose axis, as shown in Fig. 2.13. The G values can be calculated from the slopes of these lines.

Initially the primary product yield will be proportional to dose whereas the secondary product yield will be proportionate to the square of the dose.

It follows that the radiation chemical yield or G value of a product may decrease at high dose, and the concentration may ultimately reach a steady-state limit.

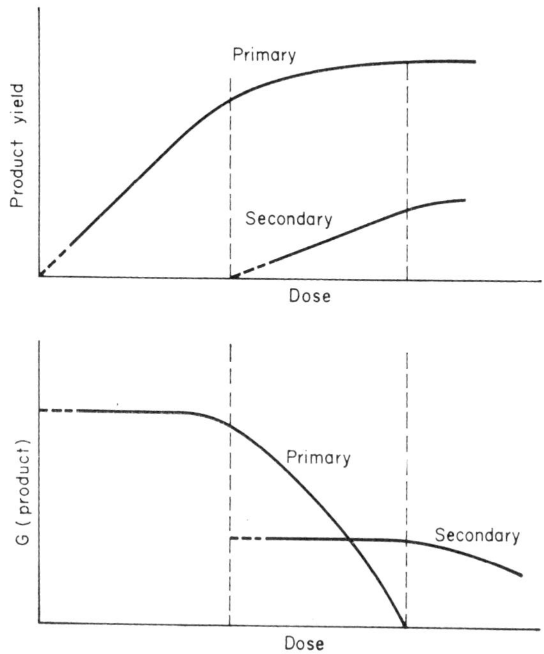

Fig. 2.13 Formation of primary and secondary products from a solute as irradiation proceeds.

Dose-rate effects

Some radicals are fairly stable and may build up to quite high concentrations. They may then be able to compete with substrate components for the primary species. This will prevent the normal series of reactions proceeding to a final product. Such reactions will show up as *dose-rate effects*, i.e. the yield of a product will vary with dose rate. Generally, a dependence of product yield on dose rate indicates a competition between second-order radical–radical reactions, the extent of which depends on the square of radical concentration, and pseudo-first-order radical–solute reactions, depending on the first power of the radical concentration.

Steady state

During irradiation at constant dose rate the primary radical species will react with substrate components or scavengers at the same rate as they are formed, and hence will reach a *steady state* concentration.

$$\left.\begin{array}{r} M \rightsquigarrow R\cdot \\ R\cdot + S \longrightarrow P \end{array}\right\} \text{equal rates} \qquad \begin{array}{r}(2.31)\\(2.32)\end{array}$$

In certain systems, usually simple ones, the products may also reach a steady state, e.g. H_2O_2 in the radiolysis of water. The overall reactions (omitting the detailed steps) are—

$$2H_2O \rightsquigarrow H_2O_2 + H_2 \qquad (2.33)$$

$$H_2O_2 + H_2 \rightsquigarrow 2H_2O \qquad (2.34)$$

The variation of steady-state concentration with conditions may give valuable information on reaction mechanisms.

Direct and indirect effects

Interaction with radiation depends on electron density. Therefore a substance of low atomic number present as a minor component of a mixture will absorb only a small proportion of the radiation. The radicals or other reactive species, which are formed when radiation acts on the major component, may, however, react with the minor component to give products. These products of *indirect* action may, or may not, be identical with the products of *direct* action of radiation on the minor components.

For example, carbon dioxide is evolved when solid benzoic acid, or its aqueous solution, is irradiated, but salicylic acid is formed only in the solution—as a result of hydroxyl radical attack. The hydroxyl radicals are formed by the action of radiation on the water.

Energy transfer

Sometimes energy absorbed at one location on a large molecule appears to be transferred to a more susceptible site elsewhere in the molecule. Thus one type of bond may be broken more frequently than would be calculated from the statistical distribution of electrons. Certain structures, such as aromatic rings, seem to take up energy and finally dissipate it as heat (as molecular vibrations). In this way systems containing these structures undergo less decomposition than would be expected.

In other cases some components of a mixture appear to be able to *transfer energy* to other components. The second component may be more readily decomposed and the result is a higher product yield. If the second component is an aromatic compound there may be considerably less decomposition (*protective effect*).

Radiation chemistry reactions

In general, most of the reactions which occur in radiation chemistry are ordinary, thermal chemical reactions, although some involve rather unusual chemical species. In some cases regions of excess energy—'hot' spots—may be present, and these regions can provide activation energies greater than those available thermally. However, these reactions also obey chemical laws, and there is no need to treat radiation chemistry as a domain beyond the realms of chemistry because of the vast energies available. We have seen that packets of a million electron volts are broken down into packets of something like thirty electron volts. Although this is still many times chemical bond energies, the species produced are identifiable chemical species and undergo chemical reactions according to established mechanisms.

A general outline of the main features of the primary radiation processes is given in Fig. 2.14.

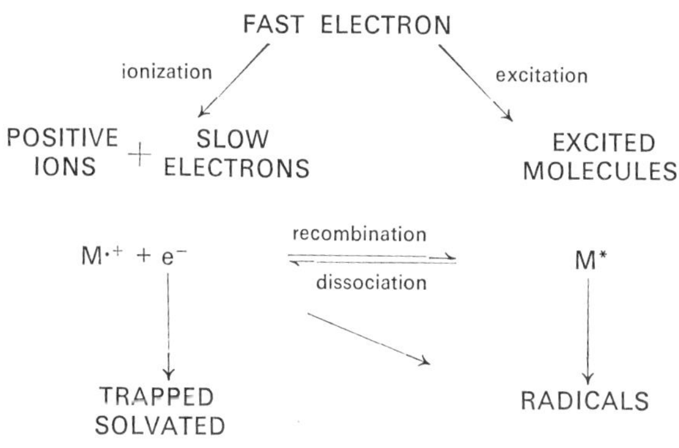

Fig. 2.14 Primary processes in action of radiation on matter.

FURTHER READING

BURNS, W. G. and BARKER, R., Dose-Rate and Linear Energy Transfer Effects in Radiation Chemistry, PORTER, G. (Ed.), *Progress in Reaction Kinetics*, Pergamon, Oxford, Vol. 3 (1965), p. 303.

† CHARLESBY, A., *Atomic Radiation and Polymers*, Pergamon, Oxford (1960).

† DAINTON, F. S., Important Early Stages in the Radiolysis of Solid and Liquid Systems, *Pure appl. Chem.*, **10**, No. 4, 395 (1965).

LEA, D. E., *Actions of Radiations on Living Cells*, Cambridge University Press (1956).

† PARSEGIAN, V. L. and CLARK, H. M., Interaction of Ionizing Radiation with Molecular Systems, in *Survey of Progress in Chemistry*, Academic, New York, Vol. 4 (1968).

PLATZMAN, R. L. *Physical and Chemical Aspects of Basic Mechanisms in Radiobiology*, National Academy of Science—National Research Council, Publ. No. 305, Washington, D.C. (1953).

Many of the general reference texts listed in Chapter 1 give accounts of the interaction of radiation with matter.

† These references should be read first.

Chapter 3
Experimental Procedures

RADIATION SOURCES

The present widespread interest in radiation chemistry has arisen as a result of the development of nuclear reactors, which can provide adequate supplies of radioactive isotopes suitable as radiation sources, and of machines which can accelerate charged particles, especially electrons, to high energies. Most investigations in radiation chemistry carried out in recent years have been with γ-rays from radioactive cobalt-60 or with electron beams from accelerating machines.

Isotopes which emit γ-radiation are the only satisfactory source of radiation which will penetrate to depths of more than about 1 cm in liquids or solids. By suitable arrangement of sources, very large samples or a number of samples can be uniformly irradiated. However, the maximum dose rate possible with an isotope source is limited by the maximum specific activity of the isotope and by the rapid decrease in dose rate with distance from the source. The maximum practical dose rate is a few Mrad/hr. Machines, on the other hand, produce beams of electrons, or other particles, and very high dose rates can be achieved (10^5 Mrad/hr steady operation, and a thousand times this rate during short pulses). The energy of the radiation can be varied, the dose rate can be altered simply by adjusting the current and usually the output can be pulsed. Energy loss as the beam passes through material is high, therefore machines can be used only to promote reactions in gases or in thin films or in the first few mm of liquids or solids.

Industrially, isotopic sources are best when a large bulk of material is to be uniformly irradiated. Machine sources are best when surface layers or thin samples are to be treated.

Isotope sources

The radioactive isotope most useful as a radiation source is cobalt-60. It is produced by neutron irradiation of normal cobalt-59 in a nuclear reactor, and high specific activities (100 curie/g upwards) can be obtained. Cobalt-60 emits two γ-rays with energies of 1.17 and 1.33 MeV (0.187 and 0.213 pJ) and has a half-life of 5.27 years. This means that it loses 1.1% of its activity each month and this rate remains constant. Therefore, once a dose-rate calibration has been made this correction can be made for the subsequent change with time.

Radioactive isotopes are used as radiation sources in three main types of physical arrangement. For small-scale experiments the radiation source can be surrounded by lead shielding and the sample introduced into close

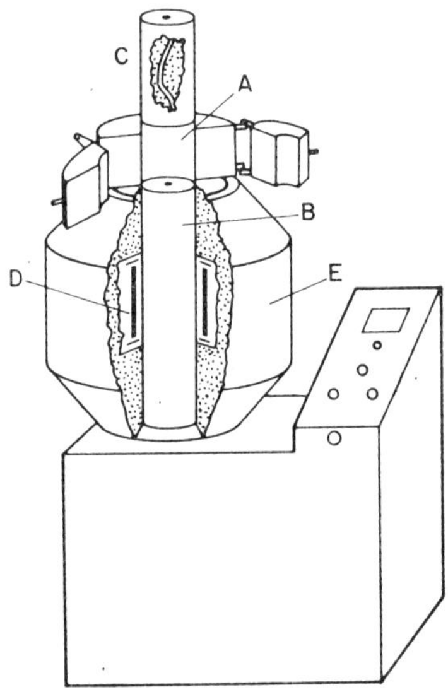

Fig. 3.1 Medium-level facility for small-scale irradiations. (A) sample cavity, (B) lower sliding shield, (C) upper shield with spiralled channel, (D) source of cobalt pencils, (E) main lead shielding.

proximity to the source by a shielded, moving drawer. A typical arrangement is shown in Fig. 3.1 and in Plate 1. A uniform radiation field is obtained by arranging a number of isotope pencils as the circumference of a cylinder. These units have the advantages of small size and high safety and can be situated in a normal laboratory. Many sources of this type are in use and they are available commercially. Their disadvantages are that (1) it is difficult to vary the radiation dose, (2) only small samples can be used and (3) operation at constant temperature is rather inconvenient (and the sample size is reduced even further). In a typical installation 1500 curies of ^{60}Co give a dose of 0.1 Mrad/hr in a 15 cm cylinder.

A much more convenient experimental arrangement is the irradiation 'cave'. The source is contained in a room or cave, usually constructed of concrete, and either the samples are moved into the cave by remote control or the source can be moved, also by remote control, into a shielded storage position. The storage position may be in a lead block or in a tank of water. Very large samples and complicated experimental apparatus, including constant-temperature baths and high-pressure equipment, can be used. The radiation dose rate can be varied by moving the sample, or altering the arrangement of the sources if a number are used. Experiments can be viewed through lead-glass windows, or by closed-circuit television, during irradiation. However, irradiation caves are expensive to build and

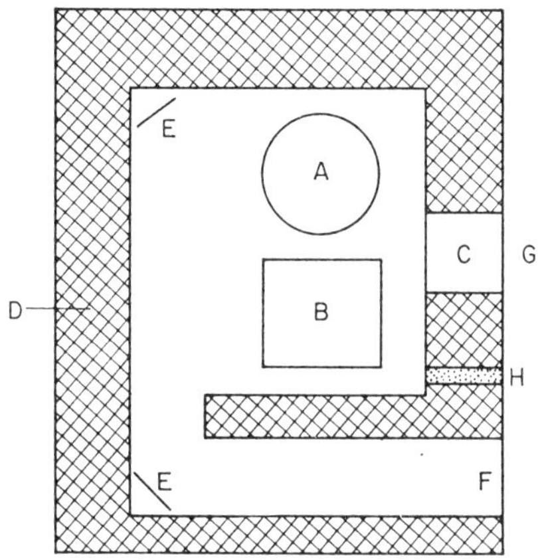

Fig. 3.2 High-level cave irradiation facility for larger-scale experiments. (A) well from which source can be raised, (B) experimental apparatus, (C) viewing window, (D) heavy shielding, (E) mirrors, (F) entrance corridor, (G) operator (H) channel for electrical leads or sampling, etc.

comprehensive safety devices are essential. A typical cave facility is illustrated in Fig. 3.2.

A source of 15000 curies of ^{60}Co in such a 'panoramic' facility will give a dose rate range of from 2 Mrad/hr for small samples 10 cm from the source to 0.005 Mrad/hr for large samples 2 metres away.

Irradiation facilities in which the radioactive material is permanently shielded have a considerable advantage in safety, especially when there are a number of operators. A version which is particularly suitable for large curie sources is the pond facility. This consists of a deep pond containing water, and the radioactive sources are stored at the bottom with the water acting as shielding. The pond is a stainless-steel tank enclosed by concrete and set in the ground. It must be watertight and contain about 10 ft (3 m) of water above the sources. The earth provides good lateral shielding, and samples are lowered into the pond in sealed cans for irradiation. However, this arrangement is unsuitable for use with complex experimental apparatus. Plate 2 shows the layout of a typical pond irradiation facility.

A plentiful source of radioactive material is 'burnt-up' or used nuclear reactor fuel. The fuel elements when they are removed from the reactor contain a variety of highly radioactive fission products and must be stored for some time until the activity decays, before they are processed to recover unused uranium or other fuel. In this period they can be conveniently used as radiation sources in pond facilities. One disadvantage is that the radiation intensity decreases very rapidly with time and frequent dose-rate measurements must be made.

There has always been some interest in high L.E.T. radiation, and for this purpose radioactive isotopes emitting α-particles have been used. Polonium-210, radium-226 and radon-222 are the most suitable, but they are very expensive.

A useful irradiation technique is self-irradiation or the use of an internal source, whereby a sample of a compound contains some isotopically-labelled material. Isotopic labelling is well established for identifying atoms, but it is also an efficient irradiation procedure, especially for radiation with low penetration, since almost all the energy absorption occurs in the sample. Tritium (hydrogen-3, β, 0.018 MeV max.), and radon-222 have been used as internal sources. Tritiated DNA has been used in radiation biological studies. Frequently self-irradiation can produce undesirable changes in isotopically labelled materials. Thus, compounds labelled with carbon-14 and iodine-131 degrade during storage, and this can have a serious effect on their properties. This may be especially important in biological compounds.

Very large isotope sources, e.g. a million curies arranged as a large flat plaque, are used in industry, particularly for sterilization. Cobalt-60 is the isotope mainly used. Low specific activity material (20–30 curie/g) is inexpensive, as cobalt rods are used to absorb excess neutron flux in nuclear reactors. Caesium-137, separated from fission products in power-reactor fuel elements, is also used as a γ-source. It requires less shielding than does cobalt and has a longer half life. There is some interest in another fission product, strontium-90, whose decay product yttrium-90 produces β-radiation.

Machine sources

The machines which produce high-energy radiation include x-ray tubes, Van de Graaff generators, linear accelerators and cyclotrons.

Much of the early work in radiation chemistry was done with x-rays, but the disadvantages include low penetration, low dose rates, a range of energies and difficulties in measuring dose rates.

Van de Graaff generators were developed for research in nuclear physics. The principle of their operation is a moving belt, which transfers charge from a low-voltage supply to an insulated storage dome, where a high potential is accumulated. This is then used to produce a beam of high-energy electrons or positively charged particles. The early machines were very large in size, in order to insulate them effectively. Modern Van de Graaff machines have the charged sphere insulated by sulphur hexafluoride or nitrogen–carbon dioxide gas under high pressure (up to 40 atm) and are relatively compact. A diagram of a Van de Graaff generator is shown in Fig. 3.3. The voltage generated is about 2–3 MeV, and the output can be continuous or pulsed.

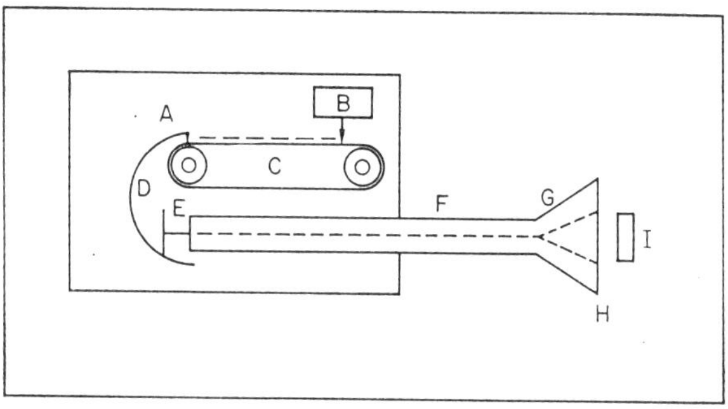

Fig. 3.3 Principal features of Van de Graaff generator. (A) insulating tank containing high-pressure gas, (B) electron injector, (C) moving belt, (D) high-voltage dome, (E) filament electron source, (F) accelerator tube, (G) scanning coil, (H) window, (I) sample or target.

The linear accelerator (Fig. 3.4) has become an important radiation source in recent years, mainly because electron beams are produced, high energies can be obtained, the energy is readily variable and the output is pulsed. It is used for industrial surface polymerization, for cancer therapy and for pulse radiolysis studies. Electrons are produced by a filament and given an initial acceleration of about 2000 volts by an electric field. They are then accelerated down a linear tube by synchronized radiofrequency waves. The acceleration increases with the length of the tube, and the electrons are separated into pulses carried by each wave.

All electron accelerator tubes must be operated under high vacuum, since attenuation in air is appreciable. Accelerators are particularly useful for investigations of gaseous systems and are essential for pulse radiolysis studies. The beam is normally about 5 mm in diameter, but it can be 'scanned' over a large area. Maximum dose rates are limited by heating of liquid and solid samples. When pulsed beams are used the dose rate during actual irradiation is much higher than the mean dose rate.

Economical high-voltage machines are now available for industrial irradiation.

Conventional power transformers may be used to provide high continuous currents at voltages up to 300 kV. This is quite adequate for irradiation of thin films at high throughputs. The major technological difficulty is in providing a thin window which will not absorb too much energy, and overheat.

Cyclotrons are widely used to provide beams of positively charged particles for nuclear physics research, and some radiation chemistry studies are done with them. The main interest is in high L.E.T. effects.

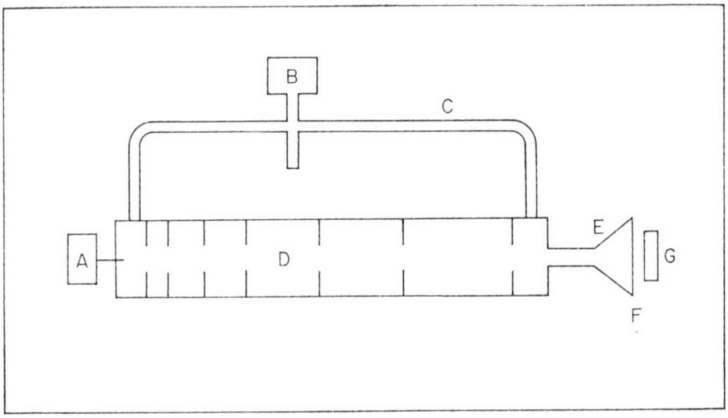

Fig. 3.4 Schematic diagram of linear accelerator. (A) electron injector, (B) klystron radiofrequency source, (C) wave guide, (D) accelerator tube with irises, (E) scanning coil, (F) window, (G) sample.

For particle-beam irradiations the sample is sometimes placed in the machine vacuum. If volatile material is present, or produced by irradiation, the sample must be placed outside the exit window of the machine behind a thin window of glass, plastic or metal.

Nuclear reactors

Nuclear reactors can be used as radiation sources, although they provide a complex mixture of radiation types and energies, which will vary with the reactor, the sample position, the age of the fuel and the operating conditions. The radiation consists mainly of neutrons, with energies ranging from fast neutrons (energy about 1 MeV) to slow neutrons (thermal energy of about 0.025 eV) and γ-rays. Fast neutrons undergo elastic scattering, producing highly energetic positive ions, especially protons from compounds containing hydrogen atoms. Slow neutrons are captured by many nuclei, resulting in nuclear transformations with associated emission of radiation, e.g.

$$^{35}_{17}\text{Cl} + ^{1}_{0}\text{n} \longrightarrow ^{36}_{17}\text{Cl} + \gamma \qquad (3.1)$$

The transformed nucleus has considerable excitation energy or recoil energy, which can result in cleavage of the bond holding the atom to the parent compound (Szilard–Chalmers effect), and this recoil atom can produce further radiation chemical effects.

A great hazard of using nuclear reactors as radiation sources is that the irradiated material may become radioactive as a result of neutron-induced nuclear transformations.

Despite the difficulties in interpretation of radiolysis resulting from such a complex mixture of radiation, there is considerable interest in the effects of nuclear reactor or pile radiation on some systems. There is continuing work on the damage done to biological material, especially living organisms. It is necessary to use a variety of materials in the construction of a nuclear reactor, and it is essential to know something about the physical and chemical changes which might affect their properties, such as strength and electrical insulation, as a result of continual irradiation. The chemical reactions induced by radiation in moderator (material used to slow down neutrons) and coolant, e.g. water or carbon dioxide, and the corrosion of fuel elements must also be studied.

The use of nuclear reactors as a direct source of radiation for industrial processes is being investigated. The fixation of atmospheric nitrogen for use in nitrogenous fertilizers is such a process. Normally, in the fission process the vast energy of the recoiling fragments is lost as heat and only the relatively small amount of γ-energy ever becomes available to produce radiation chemical reactions. In chemonuclear processes the reactants are circulated in the gaseous or liquid phase around specially designed reactor fuel elements, constructed as a honeycomb of uranium foil which is coated with a thin layer of a chemically inert metal such as platinum. The fission fragments escape from the foil, due to their high recoil energies, and very high utilization of the radiation energy is achieved. The fission energy available in the recoil energy of the fission fragments is many times that subsequently available from the radioactivity. However, there are many practical difficulties, such as decontamination of the product from the highly radioactive fission products.

DOSIMETRY

In order to determine a radiation chemical yield (G value or molecules changed per 100 eV of radiation energy absorbed) and to compare yields from one laboratory to another, or from one chemical system to another, one must know accurately the energy absorbed by the system—the dose. (100eV = 16 aJ).

Absolute dosimeters

Fundamentally the dose can be measured calorimetrically, that is, by the heat produced. In many systems, however, not all the radiation energy absorbed appears as heat, and a correction must be made for chemical energy changes. In any case, the absorption of 1 Mrad raises the temperature of a material of specific heat 1 cal (4.184 J)/g by only 2.4°C. Calorimetry is used routinely for some reactor irradiations, for electron-beam measurement and to calibrate chemical dosimeters.

There is another absolute method of measuring radiation dosage, namely by measuring the ionization produced in a particular gas under certain standard conditions. The number of ion pairs produced by the radiation is measured by the saturation current obtained when a potential is applied between two electrodes in the gas. By using tables of absorption coefficients or stopping powers, this current can be related to the dose absorbed by another system put in the place of the ionization chamber. Ion chambers are used extensively in radiotherapy.

In Chapter 2 it was shown that the exposure in roentgens is related to the number of ion-pairs produced in air. If a high voltage is applied to the electrodes of an ionization chamber containing air the number of ions produced by any given radiation field can be measured as the ions are attracted to the electrodes and discharged. Errors may arise if the walls of the chamber give an energy spectrum of secondary particles different from that given by air itself. Therefore the walls of the chamber are made of bakelite, poly(methyl methacrylate), nylon or some other air-equivalent material having mean atomic number close to that of air. One ion pair is found for each 33.7 eV of energy absorbed in air (this is the W value for the gas). It can be shown that one roentgen is equivalent to an energy absorption of 86.9 ergs (8.69 μJ) per gram in air. In practice, corrections must be applied to take account of the energy absorption characteristics of the system and of scattered and secondary electromagnetic radiation.

Alternatively, ionization measurements may be made on an ionization chamber within the material itself. Once again there are many restrictions on using this method in practice.

Ion chambers are difficult to use as absolute dosimeters because there are many restrictive conditions to be fulfilled and corrections to be applied. However, they can be calibrated so that a measure of dose rate can be readily obtained.

Because the number of ion pairs produced in a gas can be readily measured, some investigators are reverting to *M/N values*, or *ion-pair yields* for gaseous systems—a way of expressing yields which was all but discarded when it was realized that energy absorbed was the important parameter. M is the number of molecules produced or changed and N is the number of ion pairs produced for the same dose.

Internal sources

Many of the problems of dosimetry are avoided by using either radon-222 or tritium (hydrogen-3) incorporated in the system under study. All the radiation energy from the disintegration of the radioactive elements is absorbed. The dose can be calculated from the amount of isotope added.

Chemical dosimeters

Because of the practical difficulties in using either of the two absolute methods of dosimetry, radiation chemists have developed secondary standards, and usually these are chemical dosimeters. The chemical changes produced in a certain system are measured. These changes are proportional to absorbed dose.

NITROUS OXIDE The system which has gained the widest acceptance for gas phase dosimetry is pure nitrous oxide.

$$N_2O \longrightarrow N_2, O_2, NO \qquad G(-N_2O) = 12$$
$$G(N_2) = 10.2 \qquad (3.2)$$

The increase in gas pressure is measured, or for greater accuracy the unchanged nitrous oxide is frozen out and the nitrogen produced is separated and measured by gas chromatography.

FRICKE DOSIMETER The Fricke dosimeter is based on the production of ferric ion in aqueous ferrous sulphate solution on exposure to ionizing radiation. Radiation chemists, working with condensed systems (liquids or solids) almost universally relate the dose absorbed to that measured in the Fricke dosimeter. The dose absorbed is proportional to the electron density of the system, and this can be corrected for. This dosimeter has been so universally accepted that dosimetry in the gas phase is often referred to it.

Various refinements have been proposed from time to time in order to increase its accuracy, but the following recipe will give results which can be compared with most laboratories in the world.

Concentrated (95–98%) analytical reagent (A.R.) quality sulphuric acid (22 cm^3) is added carefully to distilled water ($\approx$600 cm^3)—not demineralized water. When cool, A.R. ferrous ammonium sulphate (0.56 g) and A.R. sodium chloride (0.06 g) are dissolved in the solution and the volume made up to 1 litre (1 dm^3). The solution density is 1.024 g/cm^3

and the ferrous ion concentration about 2 mM. Stored away from sunlight, the solution is stable for about one month, but should be discarded if a high blank shows the presence of ferric ion in the unirradiated solution. The optical density after irradiation is measured in a spectrophotometer at 304 mμ (nm) using 1-cm cells. The position of maximum absorption and the extinction coefficient, ε (2193 ± 6 $dm^3\ mol^{-1}\ cm^{-1}$ at 25°C is a typical value), should be found for the particular instrument being used. Molar concentration of Fe^{3+} = Optical Density/$\varepsilon[1 + 0.007(t_2 - t_1)]$ where t_1°C is the temperature at which ε is determined and t_2°C is the temperature of the optical density measurement. Alternative methods of measuring Fe^{3+} in the concentration range 0.06–0.6 mM can be used. Thus, if a spectrophotometer is not available, the ferric ion concentration can be estimated visually with a colorimeter by using a complexing agent. 'Optical density' is the same as 'decadic absorbance' so the 'extinction coefficient' is the 'molar decadic extinction coefficient'. To obtain the dose in another medium (such as pure water) a correction must be applied (see Appendix 2). The electron density per unit volume of Fricke solution relative to water is 1.022.

0.4 M sulphuric acid was chosen as being equivalent to biological tissue. Some investigators prefer to use 0.05 M sulphuric acid. This gives the same G value for ^{60}Co γ-radiation and more nearly resembles dilute aqueous solutions. Chloride ion inhibits interference from any possible organic impurity present. The explanation for this will be given in Chapter 6. G(Fe^{3+}) values for different types of radiation over the temperature range 0–65°C are given in Table 3.1. The accepted value for cobalt-60 γ-rays is 15.5. However, some authors use 15.6. It is important that the value used be stated in any report or paper.

The way that the absorbed dose can be calculated from the Fricke dosimeter measurement is shown in the Appendix.

Table 3.1. G(Fe^{3+}) values for Fricke dosimeter.

^{60}Co γ-rays	15.5
1-MeV electrons	15.5
100-kV peak x-rays	14.2
8-kV peak x-rays	13.4
Tritium (hydrogen-3) beta	12.8
2-MeV protons	8.0
^{210}Po alpha	5.1
^{235}U fission fragments	3.0

The lower limit of the Fricke dosimeter is about 4 krad, but this may be extended by using longer spectrophotometer cells, or by forming a ferric complex with a higher extinction coefficient. The upper limit for an air-saturated solution is 40 krad, as beyond that the dissolved oxygen becomes depleted. The optimum is 15–20 krad. The range can be extended to 200 krad if the solution is oxygenated and the ferrous ion concentration increased to 4 mM; $G(Fe^{3+}) = 15.5$ for cobalt-60 γ-radiation. The range can also be extended if the solution is deaerated. Then $G(Fe^{3+}) = 8.2$ for cobalt-60 γ-radiation. Another modification to the Fricke dosimeter is the addition of cupric ion immediately before irradiation, which extends the range to 10 Mrad; $G(Fe^{3+}) = 0.70$. When the sample is to be given a high radiation dose, the dose rate may be measured with the dosimeter, and the sample subsequently irradiated in the same position, the dose being proportional to the time of exposure.

For best results, the containing vessel should have glass or plastic walls at least 1 mm thick and internal dimensions greater than 8 mm. Poly(ethylene) pill packs are useful containers. Not all irradiation vessels can be designed to meet the theoretical ideal. In these cases the dosimeter solution can be irradiated in the vessel in which the sample is to be exposed. If the sample density is approximately unity, this ensures accurate dosimetry and the small correction can be made for the difference in electron density.

With care it is possible to obtain the same result on repetitive samples to $\pm 2\%$ (precision), but when all errors are taken into account the best accuracy that can be claimed is $\pm 5\%$.

The other widely accepted chemical dosimeter is the ceric dosimeter in which ceric ion is reduced to cerous ion. $G(Ce^{3+}) = 2.45$ for cobalt-60 γ-radiation. It is rather sensitive to impurities, but is useful up to several Mrad.

Other chemical dosimeters have been proposed from time to time, but none have been accepted widely, although some have advantages in specific situations.

Other dosimeters

Other dosimeters, which have been used principally for industrial installations, depend on (1) the bleaching of a red dye incorporated in poly(methyl methacrylate), (2) the darkening of a plastic, or of a coloured glass, such as one that contains 0.1% cobalt oxide, (3) the fluorescence induced in solid lithium fluoride, or in a silver-activated phosphate glass (low doses), (4) the change in colour of radiation-sensitive dyes or inks placed on packages of pharmaceuticals to show that they have received the sterilizing dose of 2.5 Mrad.

Recently, attention has been given to solid-state detectors. Although very small, these are quite effective. Moreover, the energy required to produce a charge (electron–positive hole pair) is only about one-tenth of the W value of a typical gas, hence the sensitivity is proportionally greater. The most commonly used detectors have been silicon semiconductors, suitably doped, e.g. with lithium, in the form of wafers about 1 mm thick. At present their greatest disadvantage is that large doses of radiation produce changes in the electronic characteristics.

There is a need for accurate dosimeters for use (1) in the gas phase, (2) in the low dose region <4 krad and (3) in the dose range 1–3 Mrad. Over all ranges there is a great need to be able to obtain an accuracy of $\pm 1\%$ routinely.

SAFETY

All radiation has some harmful biological effects, therefore radiation sources should be adequately shielded and personnel should always keep well away from them for as much of the time as possible. Routine laboratory work and desk work should never be carried out near a radiation source. This includes the floors above and below the source. Many countries have laws and regulations governing the use of isotopes and radiation sources and machines. At present the safe limit for the radiation dose to humans is considered to be 5 rem per year and 3 rem per 13 weeks, which compares with the annual normal background radiation of 0.1 rem, due to cosmic rays and the natural radioactivity of the earth. (The rem—roentgen equivalent man—is the dose of any radiation which produces the same biological effect as 1 roentgen of x-rays or γ-rays.)

All radiation sources should be clearly labelled with appropriate signs and symbols and personnel kept aware of the hazards involved and precautions necessary for safety. Appropriate radiation monitors in working order should always be available, and periodic measurements of the radiation levels should be made. If there is any possibility of contamination with radioactive material, hands and clothes should be monitored. Personal monitors, e.g. film badges, pocket ion chambers or solid-state detectors, should be worn by all personnel working in radiation areas. They should be appropriately positioned, e.g. on the body, wrist or fingers. However, it must be emphasized that such monitors provide only a record, not a warning.

Shielding

The thickness of shielding required for a radiation source depends on the type and energy of the radiation. Cobalt-60 γ-rays are the most penetrating radiations commonly used, and the approximate half-value thicknesses are 1 cm of lead, 5 cm of concrete and 11 cm of water. Hence adequate shielding for a 2000 curie source is provided by 1 ft (0.3 m) of lead or 10 ft (3 m) of water. In large isotope installations the inverse-square-law effect will rapidly reduce the dose rate. Care must be taken to avoid escape of radiation (shine) through ducts for electrical leads and gas vents, through tubing, over walls and around corners. Thus sample inlet drawers should be backed by further shielding and passages for leads should be spiralled. Similarly, gas venting tubes attached to irradiation cans used in pond facilities should be spiralled and hollow handling rods should be filled with water.

No person should be permitted in a cell when a machine is operating, even at the lowest power, or while a source is exposed. Lead-glass windows, mirrors or closed-circuit television can be used to view irradiations. Safety devices must be provided which prevent anyone entering a radiation field. They must be of the fail-safe type and should be tamper-proof. Also, the possibility that the mechanism for moving a source may fail while it is exposed must be considered. The radiation level should always be checked with a monitor before the cell is entered. Isotopic sources should be encapsulated and the integrity of their canning should be tested at regular intervals.

Other hazards

If samples are exposed to neutron irradiation, or γ-radiation of energy greater than 2 MeV, there is the possibility of induced radioactivity, otherwise this danger is negligible. When machines are used to produce high-energy particles, secondary x-irradiation may be produced.

Ozone is produced by irradiation of oxygen. Therefore compressed air and liquid oxygen should never be used in radiation fields. An equilibrium is established, but the boiling point of ozone (-112°C) is higher than that of oxygen (-183°C), consequently enrichment can occur after irradiation. Ozone is extremely poisonous and also an explosive oxidant for many organic materials. Liquid nitrogen condenses out oxygen from the atmosphere and this can produce ozone on irradiation. Radiation sources also produce ozone in the air during operation, and this should not be inhaled.

Irradiation of insulators, such as glass and plastics, can produce a build-up of electrical charge, and sudden discharge can cause structural break-down or ignition of inflammable gases. This is only likely to occur when high dose rates are used, e.g. with accelerators. Vacuum flasks are particularly liable to implosion as a result of discharge from the sealing-off point.

Normal chemical hazards should not be forgotten, especially as they can have unusually disastrous results. Many solvents and gases can be toxic, inflammable or explosive. Some radiation-initiated exothermic reactions can be dangerous, due to the reaction rate increasing with temperature leading to an explosion. Polymerization reactions, which have an 'induction' period, are a good example.

PURIFICATION OF MATERIALS

The experimental procedures involved in radiation chemistry studies are fundamentally the same as those used in other fields of chemistry. However, it is essential that the concentrations of some impurities be kept very low. A number of techniques are employed to achieve this.

A material can be considered pure when it contains no other substance which will interfere with the behaviour being studied. Therefore the criteria of purity are not the same for all purposes. For instance, water demineralized by ion-exchange resins might be much better than distilled water for conductivity studies. But for radiation chemical studies the minute traces of organic compounds leached out of the resins might interfere far more than would relatively large amounts of certain inert, inorganic ions.

Small amounts of impurities can interfere by competition for reactive species, or by transfer reactions whereby energy or charge passes from one molecule to another, until it is impossible to discover what the initial reactions were. The best course is to take extreme precautions in purifying materials and in keeping them pure until the products and reactions are established. One can then relax the precautions and note changes.

For example, stopcock or vacuum-tap grease is permissible for some aqueous systems at low dose, but may interfere in gaseous systems at high dose, and greaseless stopcocks must then be used. Carbon dioxide, from the atmosphere, can be significant in aqueous solution studies in the pH range 5–10, but perhaps not outside that range. Oxygen can change the nature of the products formed, and it can also inhibit some polymerization reactions. Water in organic systems will interfere with ionic reactions. Impurities seem to be very important in many gas-phase studies, e.g. the W value (eV per ion pair) for helium is profoundly influenced by small amounts of carbon dioxide. Even small quantities of impurities have a serious effect when it is necessary to use low-radiation doses to avoid complicating reactions of the radiolysis products.

The importance of very small amounts of impurities is still being discovered. A purity of 99.9% is normally considered to be high, but this represents 6×10^{20} impurity molecules per mole—more than sufficient to alter the effects of radiation.

Sometimes the effect of impurities can be countered by an additive, such as the chloride ion added to Fricke solution. In fact, the $G(Fe^{3+})$ value obtained in the absence of chloride has been used as a measure of the radiation chemical purity of water.

Most materials can be purified by the usual methods of recrystallization, fractional distillation, etc. Gases can be condensed and fractionally distilled. Solids should be sublimed, if possible, to remove occluded solvent and air. More sophisticated methods are preparative gas chromatography and spinning-band distillation.

Triply distilled water should be used in radiation chemical studies. Ordinary distilled water is refluxed with alkaline permanganate before distilling, then refluxed with acid dichromate and distilled and finally distilled in a silica still and stored in a silica vessel. An oxygen stream can be passed through the boiling water to prevent superheating and bumping.

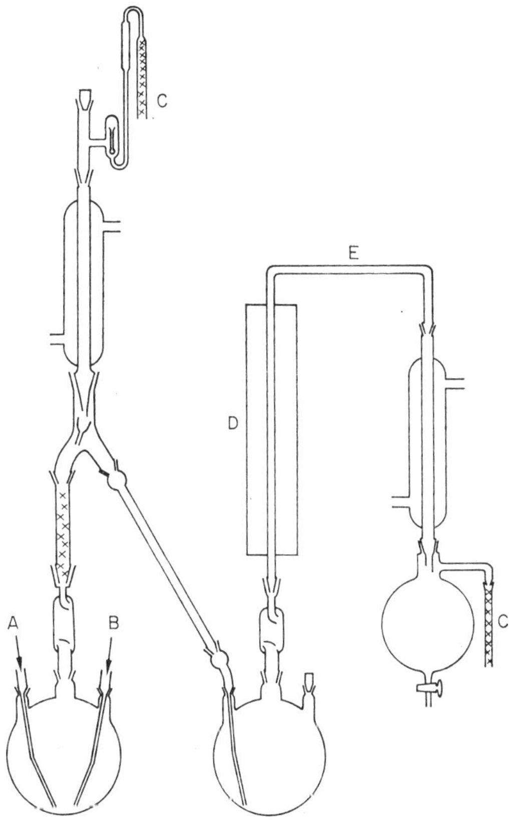

Fig. 3.5 Purification of water for radiation chemistry. (A) distilled water inlet, (B) oxygen gas inlet, (C) breathing tube packed with activated charcoal, (D) furnace at 700°C, (E) silica tubing.

As an alternative to chemical oxidation, the oxygen–water-vapour mixture can be heated in a silica tube at 700°C. Two such stages burn off any organic matter and also prevent the carry-over of any spray containing inorganic matter. An effective combination of the best features of these methods is illustrated in Fig. 3.5.

Another method of purifying water is to irradiate it. The hydrogen peroxide formed can be removed by irradiation with ultraviolet light, which has been filtered to remove those wavelengths which form hydrogen peroxide. Probably the purest water ever made was by electron-beam irradiation of water under a high pressure of hydrogen.

Storage of pure materials can be important. Solutions are best made up as concentrated stock solutions to be diluted immediately before use. Chemicals that are sensitive to light should be stored in the dark. Heat-sensitive chemicals are best kept in a refrigerator. Those sensitive to oxygen (including most unsaturated compounds) should be stored in vacuum or under nitrogen.

TREATMENT OF SAMPLES

Glass vessels used for irradiation must be carefully cleaned and detergent, etc., thoroughly removed. Chromic acid should not be used, as the last traces are difficult to wash from glass surfaces and affect many radicals. Some workers use the well-known ethanol–nitric acid mixture, but this is dangerous. It is preferable to soak vessels in concentrated nitric acid, wash several times with distilled water and finally rinse with triply distilled water. Borosilicate or silica glassware can then be heated to 500°C in a furnace to remove any remaining organic matter. This treatment is efficient and effective. Enclosed gas spaces, such as syringe barrels, must first be opened to the atmosphere by boring a hole through their end wall.

When necessary, greaseless stopcocks should be used for handling gases, and irradiation vessels provided with break-seals used to facilitate subsequent handling. Vacuum lines should be baked or flamed, if possible, and left evacuated.

Glassware, measuring apparatus and handling equipment used in the preparation of samples must be scrupulously clean and grease-free. Dispensing a measured volume of a liquid is probably more convenient, but sometimes weighing is a better way of determining sample size and is more accurate.

Bubbling an inert gas, and usually nitrogen is inert, can be used to remove most of the oxygen from an aerated liquid or solution. To prevent any volatile material from being removed from a solution, the gas should first be saturated by passing it through a large volume of the solution.

A better method of removing oxygen is the freeze–thaw method. The liquid or volatile solid in a tube attached to a vacuum line via a stopcock is frozen in liquid nitrogen and then pumped hard. The stopcock is closed and the sample allowed to warm up, permitting the trapped gases to escape. The sample is frozen again and the free gases pumped away. Four cycles are usually adequate to remove all gases. If carbon dioxide and other gases of similar boiling point are to be removed, two of the cycles should use a dry ice–alcohol (or acetone) mixture for the freezing. Finally, the tube can be sealed off under vacuum or an inert-gas atmosphere as appropriate.

Argonne syringe technique

The Argonne syringe technique has many advantages for the preparation of degassed solutions of known composition (Plate 3). Commercial hypodermic syringes are used. Luer ground-glass sockets can be purchased, but are made only to the customer's order. Standard taper cones are preferable. They can be joined via graded seals to most syringes, but it is better to seal directly to syringes made from pyrex-type glass. The solution is placed in a round-bottom flask and pumped and shaken until a water-hammer slap is heard. Inert gas (nitrogen, or better helium) is then admitted and dissolved by shaking. On re-evacuating, this gas comes out of solution, carrying with it any remnants of dissolved oxygen. Two or three cycles remove the last traces. Inert-gas pressure, without shaking, is used to lift the solution into a syringe previously filled with inert gas. The inert gas is expelled from the syringe, which is capped. The solution remains air-free for a day or more. Bubbling inert gas through a liquid in a beaker and subsequently drawing the liquid into a syringe, or simply bubbling the gas through the liquid in a syringe with the plunger removed, is less effective, but sufficient for many purposes.

Solutions of known composition can be prepared by injecting material from one syringe into another. By careful manipulation of the plungers and by expelling the drop of solution which has been in contact with air, introduction of contamination is avoided. The two can be connected by a capillary bridge with a standard socket at each end, or a thin inert needle can be made by pulling down polyethylene tubing of a size which fits tightly on the tip of the syringe. A small glass disc in the syringe is used to mix the two solutions. Solutions can be irradiated in the syringes. Syringes are particularly useful for high-pressure studies. The syringe barrel can be cut and reversed to save space.

Post-irradiation effects

Whereas most radiation chemical reactions are complete in a fraction of a second, others can continue for hours or months. In studying any system it is therefore necessary to establish that there are no post-irradiation effects leading to irreproducible results. It is often advisable to carry out separations or analyses at different times after irradiation and extrapolate to zero and infinite time. Irradiated solids frequently contain trapped radicals which are responsible for post-irradiation reactions. These reactions are minimal at low temperatures, but may be accentuated by both warming and exposure to air during separation. The radicals may be removed by melting, by chemical reaction with scavengers, or by dissolving the solid.

LIMITED FACILITIES

Limited radiation chemistry studies can be carried out without elaborate equipment. Many teachers and research workers can gain access to an x-ray set, linear accelerator or cobalt-60 source at a university, hospital or Government research laboratory. The custodians of such sources can provide dosimetry.

Many of the operations on irradiated samples are ordinary chemical operations. Generally, because low concentrations of products are desirable for mechanistic studies, spectrophotometric or chromatographic methods are used.

FURTHER READING

ATTIX, F. H., *Luminescence Dosimetry*, U.S. Atomic Energy Commission (1967).

ATTIX, F. H., ROESCH, W. C. and TOCHLIN, E. (Eds.), *Radiation Dosimetry*, 2nd edition, 3 vols, Academic, N.Y. (1966).

British Standard 4094, *Recommendation for Data on Shielding from Ionizing Radiation, pt* I *Shielding from Gamma Radiation.* British Standards Institution, London (1966).

CHARLESBY, A. (Ed.), *Radiation Sources*, Pergamon, Oxford (1964).

† FARHATAZIZ and DYNE, P. J., *A Practical Course in Radiation Chemistry*, Atomic Energy Centre, Lahore (1963).

FREGENE, A. O., Calibration of the Ferrous Sulphate Dosimeter, *Radiation Res.*, **31**, 256 (1967).

HINE, G. J. and BROWNELL, G. L., *Radiation Dosimetry*, Academic, N.Y. (1956).

JAEGER, R. G. (Ed.), *Engineering Compendium on Radiation Shielding*, 2 vols Springer-Verlag, Berlin (1968).

Solid State and Chemical Dosimetry in Medicine and Biology, International Atomic Energy Agency, Vienna (1967).

WHYTE, G. N., *The Principles of Radiation Dosimetry*, Wiley, N.Y. (1959).

The International Commission on Radiological Units and Measurements (I.C.R.U.), the International Commission on Radiological Protection (I.C.R.P.) and the U.S. National Bureau of Standards (N.B.S.) produce a series of handbooks on radiation units, health hazards and shielding. N.B.S. Handbook 85, *Physical Aspects of Irradiation* (1962), deals with dosimetry.

† These references should be read first.

Chapter 4

Special Techniques

In this chapter some advanced techniques are described. In the last few years they have contributed greatly to our understanding of the identity and reactions of the short-lived transient species formed during irradiation.

When two different types of short-lived species react with other compounds to give the same or similar products, they often react at different rates. Thus they can be distinguished kinetically, and many of the methods used in radiation chemistry are kinetic ones. It must be realized, however, that almost all the experimental and interpretative methods used in other fields of chemistry have found application in radiation chemistry.

COMPETITION KINETICS

Competition kinetic methods were used to establish that there are, in fact, two reducing species, e^- and $H\cdot$, formed in the radiolysis of water and that one of these carries unit negative charge. Similar methods have been used in other fields of chemistry to study fast reactions and unusual species.

We have seen that, for a given dose of radiation, a certain amount of each of the primary species is formed. Reaction with a scavenger (S) present in excess will continue until all the species has reacted.

$$M \rightsquigarrow R\cdot + A\cdot \qquad (4.1)$$

$$R\cdot + S_1 \xrightarrow{k_1} RS_1\cdot \longrightarrow P_1 \qquad (4.2)$$

$$-d[R\cdot]_1 = -d[S_1] = d[RS_1\cdot] = d[P_1] = k_1[R\cdot][S_1]dt \qquad (4.3)$$

The symbols have their usual significance and the units of k are volume mole^{-1} time^{-1} (usually litre mole^{-1} second^{-1} $\equiv M^{-1}s^{-1}$).

If two scavengers (S_1 and S_2) are present, then the primary species will be shared between them. This will be reflected in the relative amounts of reaction products (P_1 and P_2) obtained.

$$R\cdot + S_2 \xrightarrow{k_2} RS_2\cdot \longrightarrow P_2 \quad (4.4)$$

$$-d[R\cdot]_2 = -d[S_2] = d[RS_2\cdot] = d[P_2] = k_2[R\cdot][S_2]dt \quad (4.5)$$

The extent of reaction of the primary species with each scavenger depends on the mathematical product of reaction rate constant and concentration. A fast reaction is one with a high rate constant, but it will only proceed in preference to a slower reaction if it has a higher value for the mathematical product of the rate constant and the concentrations of the reactants.

$$\frac{d[P_1]}{d[P_2]} = \frac{G(P_1)}{G(P_2)} = \frac{k_1[S_1]}{k_2[S_2]} \quad (4.6)$$

If the relative yields of the two products are measured and the concentrations of the two scavengers are known, one can find the *relative rates*—the ratio of the two reaction rate constants for the two scavengers with that particular primary species.

A special case is that in which a primary species reacts in two or more ways with the one scavenger. S_1 and S_2 are then identical and the product yields are proportional to the relative rate constants.

$$P_1 : P_2 : P_3 \text{ etc.} = k_1 : k_2 : k_3 \text{ etc.} \quad (4.7)$$

In the general case (two competing scavengers present) it can be shown by simple algebra that

$$\frac{\Delta[R\cdot]_1}{\Delta[R\cdot]_1 + \Delta[R\cdot]_2} = \frac{G(P_1)}{G(R\cdot)} = \frac{k_1[R\cdot][S_1]\Delta t}{k_1[R\cdot][S_1]\Delta t + k_2[R\cdot][S_2]\Delta t} \quad (4.8)$$

and

$$\frac{1}{G(P_1)} = \frac{1}{G(R\cdot)}\left\{1 + \frac{k_2[S_2]}{k_1[S_1]}\right\} \quad (4.9)$$

If the reciprocal of the yield of one of the final reaction products is plotted against the ratio of the concentrations of the two scavengers, a straight line results, as shown in Fig. 4.1. The ratio (slope/intercept) is the ratio of the rate constants (k_2/k_1).

The value of this method is that only one product need be measured. The reaction rate constant for a scavenger, or a series of scavengers, relative to another scavenger, can be evaluated without measuring its product yield or, indeed, without knowing what the product is, provided that the product of the reference scavenger reaction can be measured.

This is an important method for studying the chemistry of the primary species. The results are not always unequivocal, as there are sometimes interfering reactions which upset the simple relationships. Even then, advantage may be taken of competition methods to study the interfering reactions.

In some of the older literature, mechanisms are postulated in which one molecule reacts twice with the same primary species. At doses up to several kilorads, competition considerations show that under most conditions the probability of this happening is low, but could be sufficient to upset mass balances and the determination of the yields of primary species.

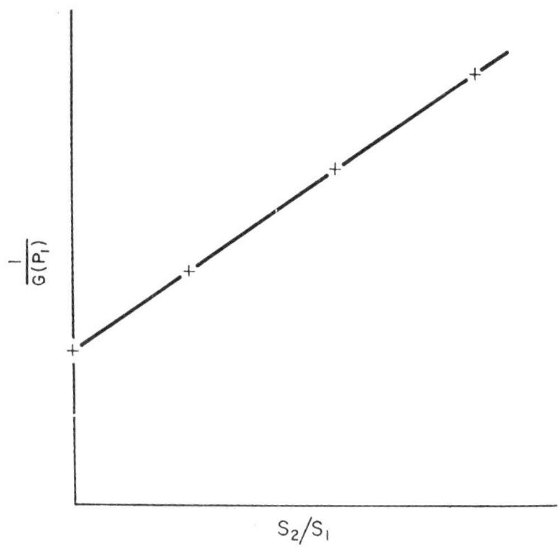

Fig. 4.1 Kinetic competition plot for reaction of a primary species with two different scavengers, S_1 and S_2.

INTERRUPTED DOSES

The technique of using a rapidly rotating sector to give pulses of light of varying duration and frequency to a system has been employed in photochemistry to obtain absolute kinetic data. It has also been used in radiation chemistry. The pulses of radiation may be obtained by using an aluminium sector in front of an electron beam or a heavy metal sector (usually lead) in front of a γ-ray beam. However, the most versatile method of obtaining interrupted doses of electron radiation is to use an accelerator which can deliver short pulses in rapid succession. In all cases there are practical difficulties in obtaining a square-wave pulse—one with an instantaneous rise and fall.

This technique can be applied only where there is a radical–radical reaction (second-order in radical concentration) competing with a radical–solute reaction (first-order in radical concentration). This can be readily detected, because there will be a dose-rate effect in continuous radiolysis.

At high dose rates, the second-order radical–radical reactions are favoured, because the radical concentrations are higher.

If the periods of irradiation and of darkness (between pulses) are long, the concentration of short-lived species, such as radicals, will build up to the steady-state value in the periods of irradiation and decay to zero in the periods of darkness.

As the pulsing rate is increased, the decay of the radicals during the dark period will not be complete, nor will the build-up of the radical concentration during the pulse reach the steady-state value. For rapid pulsing, a pseudo-stationary state will result, corresponding to continuous irradiation at an effective dose rate proportional to the fractional exposure time. It can be shown that the amount of product from radical–solute

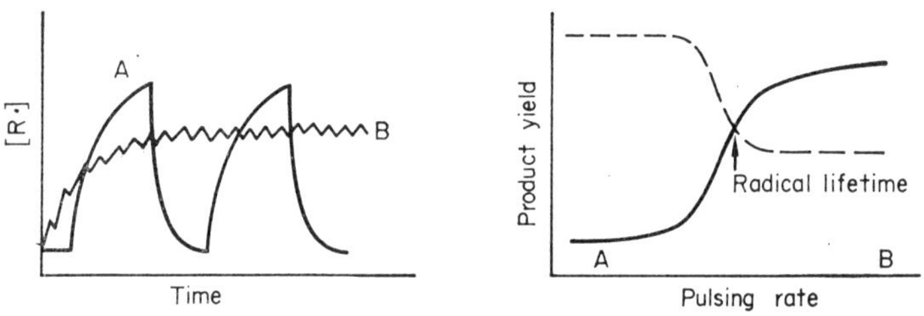

Fig. 4.2 Interrupted dose. Variation of product yields with pulsing rate. Full line—product of radical–solute reaction. Dotted line—product of radical–radical reaction. (A) Slow pulsing. (B) Rapid pulsing.

reaction will be greater for rapid pulsing, as shown in Fig. 4.2. Moreover, the rise in the product yield occurs when the pulse interval approximates to the lifetime of the short-lived species. From this, the rate of the reaction can be evaluated. The theory is quite complex and in practice, a mathematical model is set up to follow the yield curve and give accurate rate data.

FLOW METHODS

Another method of investigating short-lived transient species is to irradiate a stream of liquid or gas and to examine this at various positions downstream by such methods as optical absorption. Under constant-flow conditions, a steady state is produced and the distance down-stream is a function of time. Thus the appearance or disappearance of species with time can be followed. Furthermore, other reagents can be mixed with the stream after irradiation and, in this way, the chemical reactions of the species followed. Figure 4.3 is a schematic diagram of the experimental arrangement.

If excess scavenger is added at the mixing point, the resulting solution can be analysed at leisure, and hence the concentration of short-lived species at the point and time of mixing can be found.

Using flow methods it was possible to show that the pK of $HO_2\cdot$ in aqueous solution was about 4.4.

$$HO_2\cdot \longrightarrow H^+ + O_2\cdot^- \quad pK\ 4.4 \qquad (4.10)$$

Later, other techniques have given a value of 4.5 in good agreement with this.

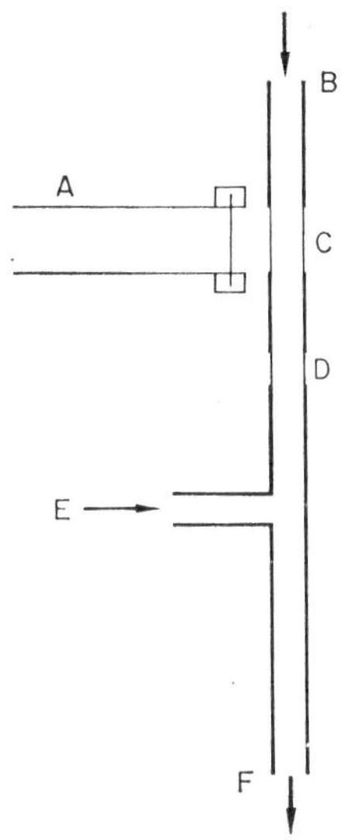

Fig. 4.3 Flow system. (A) Accelerator. (B) Solution inlet. (C) Irradiation cell. (D) Spectrometer. (E) Addition of reagent. (F) to Analysis.

ELECTRON-SPIN RESONANCE

The development of *electron-spin resonance* (e.s.r.), also known as electron paramagnetic resonance (e.p.r.), has been particularly valuable in the study of radicals in radiation chemistry. Electron-spin resonance detects the paramagnetism (spin) of unpaired electrons, at concentrations down to about 10^{-8} M under suitable conditions.

The basic principle is that electron spin is quantized into two energy levels in a magnetic field. Transitions from the lower to the upper level can be induced by microwave radiation with energy equal to the difference between the levels, resulting in resonance absorption. The intensity of absorption is proportional to the difference between the number of spins in the two levels and it increases greatly at low temperatures. At any temperature it is proportional to the total number of spins.

Electron-spin resonance spectrometers were originally constructed using 3 cm wavelength klystrons developed for wartime radar work and corresponding magnetic fields of 10 000 gauss, and these X-band instruments are now standard. There are advantages in obtaining measurements at a second wavelength, and instruments operating in the K- and Q-bands are occasionally used.

The sample is placed in a cavity in the wave guide from the klystron and situated between the pole faces of an electromagnet. It is irradiated with microwaves of a fixed frequency, which has been tuned to resonance with the cavity, and the magnetic field is varied. A modulated sweep is

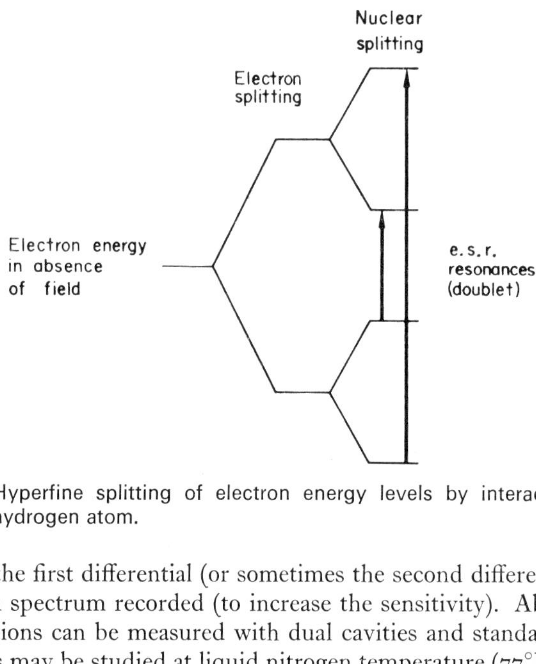

Fig. 4.4 Hyperfine splitting of electron energy levels by interaction with a hydrogen atom.

used and the first differential (or sometimes the second differential) of the absorption spectrum recorded (to increase the sensitivity). Absolute spin concentrations can be measured with dual cavities and standard samples.

Samples may be studied at liquid nitrogen temperature (77°K) or liquid helium temperature (4°K) by inserting a special vacuum flask into the cavity, or over a temperature range from about −180°C to +200°C by passing cold nitrogen vapour through a heater and into the cavity. Electron spin resonance spectra may be recorded during ultraviolet irradiation by using a cavity with a silica window. A special cell of small volume is required for aqueous solutions, as water absorbs microwave radiation. A typical e.s.r. spectrometer is shown in Plate 4.

Electron-spin resonance would be a technique of very limited value if it could be used only to detect unpaired electrons, but it is also possible to identify free radicals and in many cases characterize their structure.

Interaction of the electron with paramagnetic nuclei, e.g. hydrogen, results in hyperfine splitting of the electron energy levels. This produces multiple resonance lines. The number, positions and intensities of the lines are characteristic of the radical. Each of the two energy levels of the electron is split into a doublet by interaction with a hydrogen nuclear spin. This results in a doublet e.s.r. spectrum for an isolated hydrogen atom, as shown in Fig. 4.4. The splitting in this case is $\gtrsim$510 gauss, but it is much less for interaction with hydrogen atoms in a radical (1 gauss = 2·80 MHz).

Hyperfine splitting by hydrogen atoms has been the main reason for the wide use of e.s.r., although there has been some work with nitrogen, fluorine, carbon-13 and deuterium.

Detailed investigations have been made into the relationship between

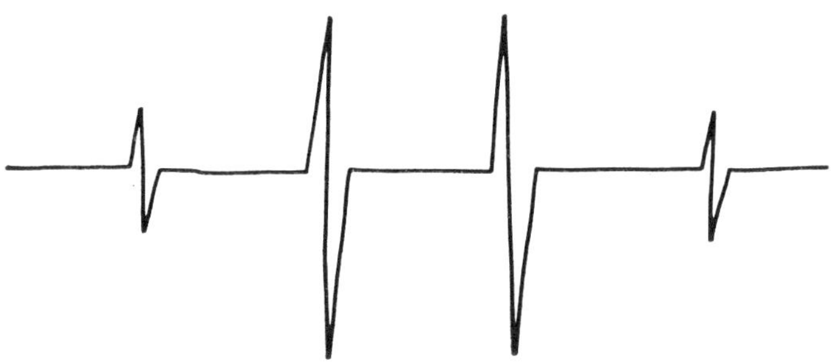

Fig. 4.5 Electron-spin-resonance spectrum of methyl radical in electron-irradiated methane.

radical structure and e.s.r. spectra. There has been great interest in the evidence which e.s.r. provides for theoretical models of electronic structure, especially the electron densities on certain atoms.

Intermediate species in radiolysis which have been observed by e.s.r. may be divided into five categories—(1) electrons, (2) hydrogen atoms, (3) inorganic radicals, (4) aliphatic radicals and (5) aromatic radicals. The radicals may be neutral, anionic or cationic.

The e.s.r. spectra of aromatic radicals contain a great number of resonance lines, but can be interpreted by assuming that the unpaired electron occupies a molecular π-orbital and interacts with all hydrogen atoms α or β to the ring carbons, i.e. attached to them, or one atom removed in a side group or side chain.

In aliphatic radicals the unpaired electron is considered to occupy a p-orbital. Interaction is effectively restricted to α- and β-hydrogen atoms, although a small interaction with γ-hydrogen atoms has been observed.

Liquid-phase studies give well-resolved spectra for radicals, but it is difficult to obtain adequate radical concentrations by irradiation. A technique which has been used is to direct an electron beam on the sample in the cavity through an axial hole in the magnet. The methyl radical spectrum observed in electron-irradiated liquid methane is shown in Fig. 4.5. The spectra of methyl and ethyl radicals have been similarly observed in irradiated ethane.

Alternatively, the radicals can be trapped in solid samples irradiated at low temperatures. The resonance lines are then very broad and difficult to resolve. However, an advantage is that the reactions of the radicals can be followed as the sample is warmed up.

Electron-spin resonance studies have now been made of a large number of radicals, and collections of spectra have been published.

MATRIX ISOLATION

A major problem in studying intermediate species in radiolysis reactions is that their lifetimes are very short. Consequently, the steady-state concentrations are usually too low for optical or e.s.r. spectroscopic examination. One method of overcoming this problem is *matrix isolation.* This involves trapping the species in a medium with which they do not react and through which they do not diffuse. Frequently this is best achieved at low temperatures. Sometimes annealing, heat treatment, or irradiation with light (photobleaching), simplifies the situation by reducing the number of species present.

Electron-spin resonance spectra of solids are not nearly as well resolved as those of liquids, due to the much broader resonance lines. The lines are broader as a result of less efficient relaxation (return from excited to ground state) and because anisotropic effects do not average to zero in solids, as in liquids. Greater resolution can often be achieved by warming the sample. Matrices of certain 'round' molecules, such as SF_6, do not hinder the free rotation of isolated solutes. Consequently the e.s.r. spectra are almost as well resolved as in the liquid phase.

Both atoms and radicals have been observed by e.s.r. in irradiated solids at low temperatures. Thus, the e.s.r. spectrum of γ-irradiated methane at 4°K contains a doublet due to hydrogen atoms and a quartet due to methyl radicals, as shown in Fig. 4.6. Note the difference between the line widths of the methyl radical spectrum in the solid and liquid phases (cf. Fig. 4.5).

Electrons trapped in anion vacancies (F-centres) give singlet e.s.r. spectra in irradiated alkyl halides. When many organic glasses are irradiated they develop a deep colouration and a single e.s.r. line. This is attributed to an electron, and readily bleached by light. Similar results have been obtained with aqueous NaOH glasses at 77°K.

Hydrogen atoms are normally very reactive, but they have been observed at 4°K in irradiated hydrogen and methane (Fig. 4.6). Their characteristic doublet is often found in irradiated organic compounds at 77°K, but the

Plate 1 Isotope facility for small-scale irradiations (p. 32).

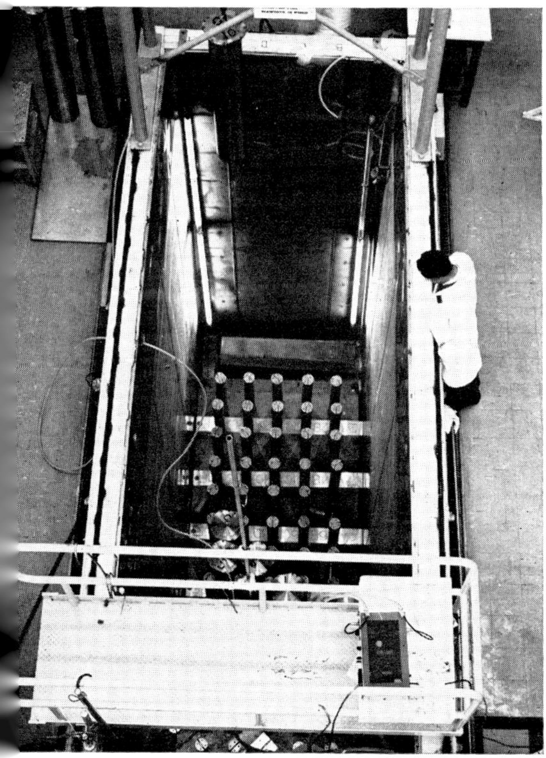

Plate 2 Pond-irradiation facility using spent fuel elements and cobalt-60 (p. 34).

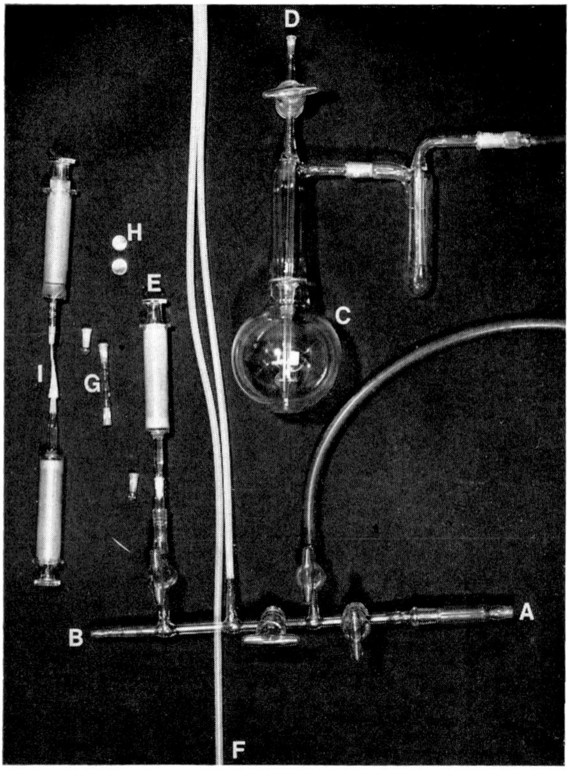

Plate 3 Argonne syringe technique. **A** to vacuum line, **B** to inert-gas cylinder, **C** degassing flask, **D** syringe is attached here for filling with solution, **E** filling with gas, **F** gas overflow through mercury seal, which also indicates pressure less than atmospheric, **G** capillary transfer between syringes, **H** mixing plates, **I** transfer through polyethylene needle (p. 47).

Plate 4 Electron spin resonance spectrometer in operation (p. 60).

concentration is always low. Probably the hydrogen atoms are absorbed on the surface of the silica sample tube. Stable hydrogen atoms have been formed at room temperature by irradiation of silica gel. They are also stabilized by basic anions, presumably by hydrogen bonding, and have been trapped in irradiated oxy-acids such as $HClO_4$, H_2SO_4 and H_3PO_4 in the solid state at low temperatures.

Radicals produced in the gas phase can be condensed on to cold surfaces together with an inert diluent, e.g. argon or nitrogen, and examined. Radicals can also be trapped at room temperature in suitable matrices, e.g. ketyl radicals have been trapped in poly(methyl methacrylate).

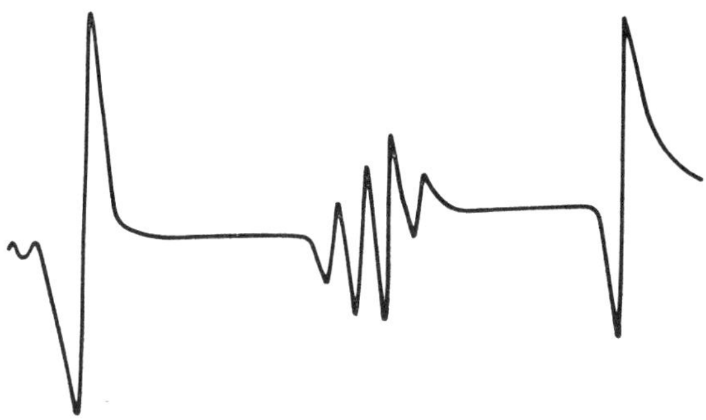

Fig. 4.6 Electron-spin-resonance spectrum of γ-irradiated solid methane at 4°K showing hydrogen atoms and methyl radicals.

Electron-spin resonance has produced the major discovery that radicals produced in irradiated single crystals are usually oriented in the crystal lattice. This has greatly aided interpretation of the e.s.r. spectra of solids. It has enabled determination of the anisotropy of both the position of the spectrum (g-factor) and the hyperfine splitting by measurement of the spectrum at different angles of rotation of the crystal, relative to the magnetic field.

The reactions of radicals trapped in solid matrices may also be observed by e.s.r. Considerable work has been done on the disappearance of radicals, and this has been correlated with mobility, especially near phase changes. Kinetic studies on radical–radical reactions have not been very useful because the diffusion rate is the controlling parameter. Studies of radical–molecule reactions may be more fruitful.

Optical spectroscopic measurements in the visible and ultraviolet regions have also been made on frozen systems. However, they do not give as much information concerning radical structures as do e.s.r. measurements.

PULSE RADIOLYSIS

It is possible to follow the appearance and disappearance of transient species by means of pulse radiolysis and hence to measure absolute reaction rate constants. Pulse radiolysis does for radiation chemistry what flash photolysis does for photochemistry, and events down to the nanosecond (10^{-9} s) region can be followed.

The system under study is subjected to an intense burst of ionizing radiation for a very short time—microseconds (10^{-6} s) to nanoseconds (10^{-9} s). An electron accelerator such as a linear accelerator or Van de Graaff may be used to provide such a pulse. Alternatively, the field-emission principle may be utilized by applying a strong electric field to draw a very large current of electrons (some thousands of amperes) from an electron source during a few nanoseconds.

The ensuing reactions can then be followed during the pulse or over a further interval of time with detectors capable of very rapid response times. A number of transient species, usually radical species, ions or excited states, have been identified and characterized.

Experimental

A short pulse of ionizing radiation, usually from an electron accelerator, is delivered to the system under investigation. The system can be solid, liquid or gaseous, or a solution. The experimental arrangements may be different in each case. Any means can be used to detect the species formed,

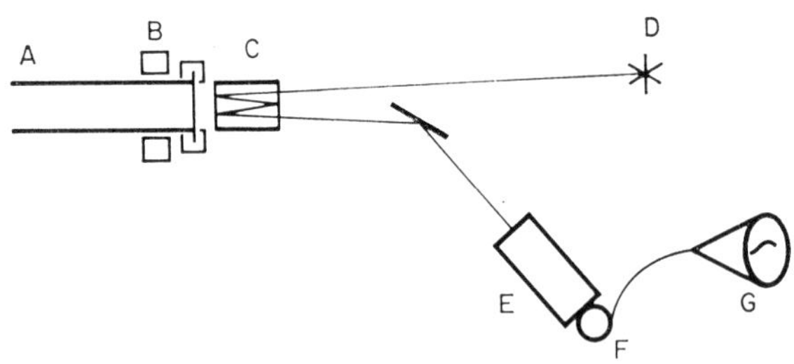

Fig. 4.7 Pulse radiolysis. (A) Accelerator. (B) Sensor. (C) Irradiation cell. (D) Light source. (E) Monochromator. (F) Photomultiplier. (G) Oscilloscope.

provided it is sensitive to small concentrations or small changes in concentration, and its time of response is short enough to follow the changes. Optical absorption has been the most widely used. The light source is usually a high-pressure xenon continuum, but lasers have also been used. The experimental set-up is shown in Fig. 4.7. Conductivity and polarography have also been used. Suitable fast infrared detectors are now available. Electron-spin resonance can be used to follow free radicals.

Normally, the scan time is too long to take complete spectra giving information on the structure of short-lived species. However, if successive pulses are used, it is possible to accumulate signals and thus build up a composite picture.

A grating and photographic plate may be used to obtain a complete optical spectrum from a single pulse. For kinetic investigations, however, the wavelength is selected by a monochromator and the light intensity measured by a photomultiplier connected to a cathode-ray tube (oscilloscope) with a fast time sweep. The trace can be photographed for later measurement. The vertical displacement on the screen is a function of the concentration of the absorbing species.

Each pulse of the electron beam can be monitored by a sensing coil or an ionization chamber, giving a signal on an oscilloscope. The position of the pulse along the trace can be obtained from the Cerenkov light emission.

Interpretation

An idealized curve is given in Fig. 4.8 to illustrate the sort of information that can be obtained from a trace following a pulse lasting a fraction of a microsecond.

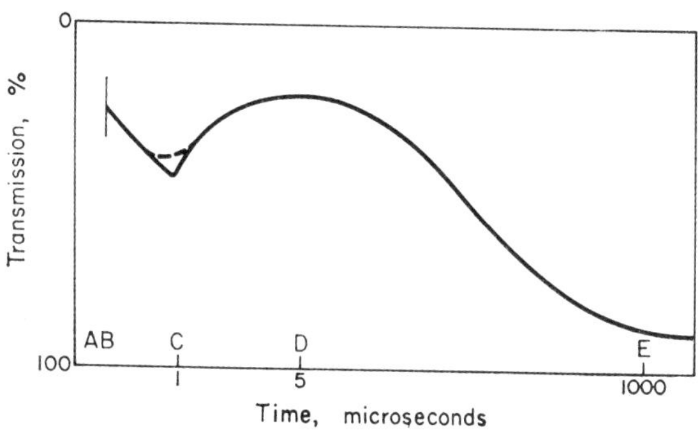

Fig. 4.8 Idealized trace. AB—pulse and Cerenkov radiation. BC—primary radicals disappear. CD—radical–solute adducts appear. DE—adducts disappear.

In the region *AB* the electron pulse is being delivered. There is a considerable amount of Cerenkov radiation produced, but some observations can be made (at wavelengths outside the Cerenkov region).

$$\mathrm{M} \rightsquigarrow \mathrm{M}^{\cdot +} + \mathrm{e}^- \qquad (4.11)$$

$$\mathrm{M} \rightsquigarrow \mathrm{M}^* \rightarrow \mathrm{R}\cdot + \mathrm{R}'\cdot \qquad (4.12)$$

At the point *B*, the primary species are present and can be observed by suitable detectors. Over the region *BC*, the primary species are disappearing by reaction with scavengers. If the product does not interfere, it is possible to measure the rate of disappearance of the primary species and from this to calculate the absolute rate constant.

$$R\cdot + S \longrightarrow RS\cdot \tag{4.13}$$

$$-\frac{d[R\cdot]}{dt} = k[R\cdot][S] \tag{4.14}$$

$$k = \frac{\ln([R\cdot]_0/[R\cdot]_t)}{[S]t} \tag{4.15}$$

where $[R\cdot]_0/[R\cdot]_t$ is the ratio of primary species concentrations at point B and at time t. In practice, something proportional to concentration, such as optical density (absorbance), is used.

The region *CD* can represent the continued growth of the product of reaction *4.13*. Since $-d[R\cdot] = d[RS\cdot]$, the rate of growth of the immediate product can be used under certain conditions, in an equation similar to equation *4.15*, to measure absolute reaction rate constant.

Alternatively, the region *CD* can represent the growth of further products from reaction with other solutes.

$$RS\cdot + X \longrightarrow RSX\cdot \tag{4.16}$$

Over the region *DE* the species dimerize, decompose, disproportionate or otherwise react to give metastable or final stable products.

$$RS\cdot \longrightarrow P \tag{4.17}$$

At point E, i.e. after a millisecond, the only absorption will be that due to relatively long-lived products.

By choice of conditions, it is possible to enhance or eliminate various sections of the idealized trace of Fig. 4.8, and to obtain optimum conditions for the measurement of particular effects. An experimental trace is given in Plate 5.

Dosimetry

The use of intense, ultra-short pulses introduces special problems of dosimetry. The total dose delivered in a microsecond pulse is not high, of the order of 1000 rad, but the dose rate is extremely high. At dose rates greater than about 5×10^8 rad/s, the Fricke dosimeter must be modified by omitting the chloride ion and saturating the solution with oxygen. Under these conditions, $G(Fe^{3+})$ is still 15.5 at much higher dose rates. The high optical absorption of the hydrated electron may be used as a convenient dosimeter. The value of $G \times \varepsilon$ (ε = extinction coefficient) has been determined accurately for the hydrated electron in pure water at various wavelengths. $G \times \varepsilon$ for other species can be evaluated from their optical densities after an identical pulse. If ε is known, then the value of G can be obtained.

Aqueous systems

The most extensive studies using pulse radiolysis techniques have been made on aqueous systems. The presence of the hydrated electron was demonstrated in irradiated alkaline water. The absorption spectrum is shown in Fig. 4.9. The extinction coefficient of the hydrated electron is so high that its disappearance, as it reacts with a scavenger, can be readily measured.

$$e^-_{aq} + S \longrightarrow S\cdot^- \qquad (4.18)$$

By using equation *4.15*, absolute rate constants have been evaluated for several hundred solutes, chosen from all classes of compounds. Patterns of behaviour are emerging and we now have a much better understanding of these reactions.

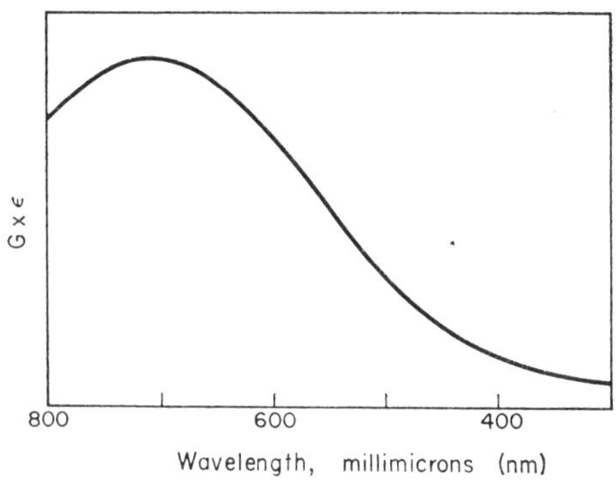

Fig. 4.9 Absorption spectrum of hydrated electron.

Since the hydroxyl radical absorbs only weakly in the far ultraviolet region, rates of reaction cannot be determined from the rate of its disappearance. Instead, the rate of appearance of intermediate products (region *CD* of Fig. 4.8) has been used to evaluate absolute reaction rates for the hydroxyl radical, using an equation of the form of equation 4.15.

A second scavenger will slow down the appearance of the first immediate product, so a combination of pulse radiolysis and competition methods may be used to extend measurements to a wider range of scavengers.

The spectra of a large number of electron–solute adducts and hydroxyl radical–solute adducts have been obtained, and the subsequent reactions of some of these have been followed. A common reaction of electron adducts in protic solvents is rapid protonation.

$$S\cdot^- + HA \longrightarrow HS\cdot + A^- \qquad (4.19)$$

Gases

For pulse radiolysis studies of gaseous systems the problem is to produce a high enough concentration of species to study. The best conditions are given by pulses of relatively low voltages ($\approx$ 200 kV) and very high currents (thousands of amperes). Such electron beams can be produced by accelerators incorporating exploding metal filaments and similar devices. Somewhat different techniques for identifying and measuring the transient species may have to be used. Microwave resonant frequency changes can be used to determine electron concentration.

Organic systems

In organic systems it has been possible to observe the formation and decay of excited triplet states and the transfer of excitation from one molecule to another. Also, the formation of anions by capture of electrons by neutral molecules has been observed. At very short times, of the order of nanoseconds, parent positive ions are present.

The reactions of these species and of free radicals can be followed. Organic systems have not been studied as intensively as have aqueous solutions, and there is much still to be done to obtain a more complete understanding of the reactions involved.

Monomers and polymers

Pulse radiolysis techniques are also helping to sort out the reactions which occur during radiation-induced polymerization of monomers. The initiating species and the growing chains can be identified.

'Protection' of polymers in aqueous solution, for example by sulphur compounds, has been investigated, and also the species formed in irradiated solids at ambient temperature.

THEORY AND CALCULATIONS

Considerable progress has been made in placing radiation chemical studies on a firm theoretical basis. A major difficulty in the development of quantitative theories has been the impossibility of separating the physical aspects from the chemical aspects. Both play a role in the formation of the products, which must be identified and measured experimentally in order to test the validity of the theories. The availability of pulse radiolysis facilities at the nanosecond timescale has provided data of more value for this purpose and digital computers have permitted more rigorous treatment of theories.

By making a few assumptions, which are based on experimental evidence from other fields, about the interaction of moving particles with the constituents of the medium through which they are passing, and about the interaction between charged species, it is possible to develop mathematically a reasonable picture of the physical events which take place and of their spatial distribution in the medium. This picture has been described in Chapter 2.

This theoretical treatment has been expanded and modified over the years, but it is never possible to be certain that all the relevant factors have been taken into account. Nevertheless, such mathematical studies are invaluable for indicating the sequence of events, and for providing ideas from which new and crucial experiments can be devised.

The primary species formed close together in blobs and short tracks are able to react with their immediate neighbours, or diffuse into the bulk of the solution. Mathematical diffusion theory has been used to compare the proportion which should escape with that observed to react.

The dielectric constant is a very important parameter in determining the probability of an electron escaping from its geminate partner in an isolated spur. This probability is high in a liquid of high dielectric constant, such as water, and low in a liquid of low dielectric constant.

The species that escape intra-spur reaction and diffuse into the bulk of the solution can be considered to be distributed homogeneously. Ordinary chemical kinetics can be applied to their reactions. Computer simulation has been used to follow the interactions of primary species with each other and with scavengers at reaction rates which are known or assumed. The growth of products calculated in this way by analogue methods, by integrated rate equations or by stepwise digital computation can then be compared with pulse radiolysis results, or with end-product analysis. Thus, postulated reactions and assumed rate constants can be tested.

Finally, much of the tedium of digesting pulse radiolysis results has been removed by using electronic methods to convert oscilloscope traces into numerical data, i.e. analogue–digital conversion, and by processing the data with computers.

FURTHER READING

AYSCOUGH, P. B., *Electron Spin Resonance in Chemistry*, Methuen, London (1967).

BIELSKI, B. H. J. and GEBICKI, J. M., *Atlas of Electron Spin Resonance Spectra*, Academic, New York (1967).

† DORFMAN, L. M. and MATHESON, M. S., Pulse Radiolysis, in *Progress in Reaction Kinetics*, PORTER, G. (Ed.), Pergamon, Oxford, Vol. 3, p. 237 (1965).

EBERT, M., KEENE, J. P., SWALLOW, A. J. and BAXENDALE, J. H. (Eds.), *Pulse Radiolysis*, Academic, London (1965).

FISCHER, H., *Magnetic Properties of Free Radicals*, in *Landolt-Bornstein*, New Series, Group II, Vol. 1, Springer-Verlag, Berlin (1965).

MATHESON, M. S. and DORFMAN, L. M., *Pulse Radiolysis*, M.I.T. Press, Cambridge, Mass. (1969).

† NORMAN, R. O. C. and GILBERT, B. C., Electron Spin Resonance Studies of Short-lived Organic Radicals, in *Advances in Physical Organic Chemistry*, GOLD, V. (Ed.), Academic, New York, Vol. 5, p. 53 (1967).

PIKAEV, A. K. (HART, E. J., Ed.), *Pulse Radiolysis of Water and Aqueous Solutions*, Indiana University Press, Bloomington and London (1967).

† SYMONS, M. C. R., The Identification of Organic Free Radicals by Electron Spin Resonance, in *Advances in Physical Organic Chemistry*, GOLD, V. (Ed.), Academic, New York, Vol. 1, p. 284 (1963).

† WHELAN, D. J., Intermediates in Radiation Chemistry at Low Temperatures, *Chem. Rev.*, **69**, 179 (1969).

† These references should be read first.

Chapter 5

Gaseous Systems

The passage of high-energy, ionizing radiation through gases produces ionic, excited and free radical species, as in other phases.

$$M \rightsquigarrow M\cdot^{+}, e^{-}, M^{*}, A\cdot, B\cdot \tag{5.1}$$

These species then interact to give the final radiolytic products, which can be separated and measured.

EARLY STUDIES

Because ionization was such an obvious phenomenon following the passage of high-energy radiation through a gas, most early mechanisms of radiolysis were based on the formation of ions and their neutralization. Often, the total yield of products was greater than the number of ion pairs. This was explained by postulating that neutral molecules formed *clusters* around the ions, in much the same way as water molecules, being polar, will condense on ions in a cloud chamber, so that on neutralization a number of molecules were changed.

Then it was realized that the observed reactions could be simply explained on the basis of formation of excited species, which could give rise to free radicals and chain reactions involving a series of radicals. Many years later, it was found that there were also many important reactions involving ions.

It is now realized that both ions and excited molecules are formed, and that these may decompose, or react with other entities, or transfer energy to other entities. Recently, the neglected *ion-cluster* concept has been revived and there is little doubt that clusters are formed in some systems.

ION CURRENT

In gases, the ionization produced may be measured readily. If an electrical potential is applied between two electrodes inserted in the gas, a current flows, as in an ionization chamber (Fig. 5.1). As the potential is increased, more ions are attracted to the electrodes and the current increases proportionately, until a *saturation current* is reached (Fig. 5.2). At that stage all ions produced are neutralized at the electrodes and any further increase in potential produces no further increase in current (until secondary ionization occurs).

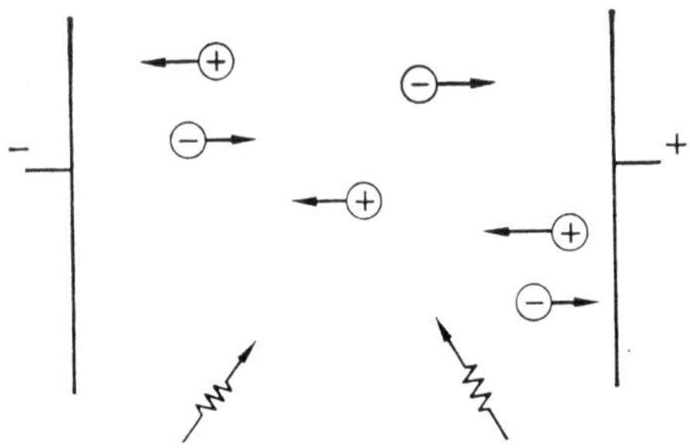

Fig. 5.1 Effect of an electric potential field on a gas under irradiation.

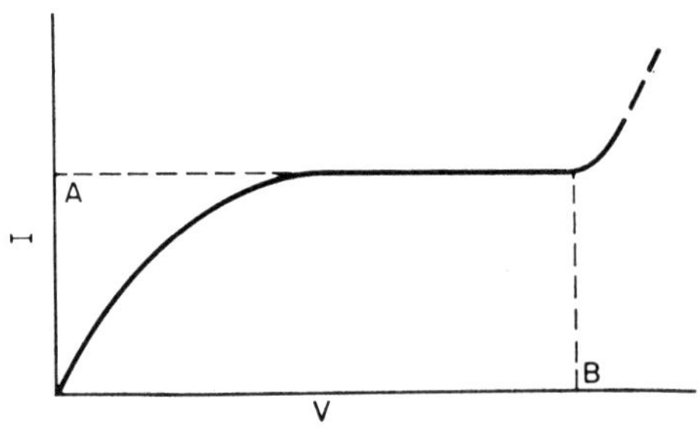

Fig. 5.2 Current (I) versus voltage (V) curve for a gas under irradiation. (A) saturation current. (B) onset of secondary ionization.

Because ion pairs can be separated by an electric field in this way, and can be counted, gases are attractive systems for fundamental studies.

W value

The *W value*—the amount of energy in electron volts absorbed in the gas for each ion-pair found—has been measured for most gases. It usually lies between the limits of 20 and 45 eV. About half of this absorbed energy is required for the actual ionization. Energy lower than the *ionization potential*, the energy required to remove an electron (usually about 9–16 eV, i.e. 1.5–2.5 aJ), may raise the molecule to an *electronically excited state*, without the electron escaping. There are many such states in a complex molecule. Usually the lowest is the most important in radiation chemistry, because decomposition takes place from this state. So energy not used in ionization may produce excitation of the molecules and of the positive ions. A small amount goes into *sub-excitation electrons*.

A measure of the amount of energy available for excitation can be obtained by comparing the W value with I, the ionization potential of the molecule. Typical W and I values are given in Table 5.1. For these substances about 40–60% of the absorbed energy is not accounted for by ionization. The W values listed are for electron irradiation. They are slightly different for α-particle irradiation. The W value for helium is very sensitive to impurities because energy can be transferred from the excited helium atoms to the impurity molecules, which may then be ionized. Thus, the W values for mixtures of gases must be measured, and

Table 5.1. *W* values and ionization potentials (*I*) for some typical gases.

Gas	W		I		W−I	
	eV	aJ	eV	aJ	eV	aJ
Oxygen	30.8	4.93	12.1	1.94	18.7	2.99
Nitrogen	34.6	5.54	15.6	2.50	19.0	3.04
Argon	26.2	4.19	15.8	2.53	10.4	1.66
Helium	41.5	6.64	24.6	3.94	16.9	2.70

not calculated from the composition of the mixture. In gases, the values of G_e, the number of electrons formed per 100 eV of radiation energy absorbed, which must be equal to $100/W$, lie between the limits of 2.2 and 5. In most cases the values are between 3 and 4. The product yields will be given by $G(\text{product}) = 100/W \times M/N$.

Energy transfer

Excitation energy can be transferred from one component to another. If the ionization potential of the second component is lower than that of the first, it will be preferentially ionized, although it may be only a minor component, or even a trace of impurity. For example, a trace of argon in helium will give additional ion pairs, and hence a W value lower than that for pure helium.

$$\mathrm{He}^* + \mathrm{Ar} \longrightarrow \mathrm{He} + \mathrm{Ar}^{\cdot+} + \mathrm{e}^- \qquad (5.2)$$

On the other hand, if argon is present in hexafluoroethane, energy transfer takes place from the argon to the C_2F_6. Also, energy transfer from the argon ion to methane, producing dissociative ionization of the methane, has been shown to occur.

$$\mathrm{Ar}^{\cdot+} + \mathrm{CH_4} \longrightarrow \mathrm{Ar} + \mathrm{CH_3}^+ + \mathrm{H}\cdot \qquad (5.3)$$

COMPARISON WITH CONDENSED SYSTEMS

Because of the low density of gases, there is little absorption of energy from γ-radiation. For this reason much of the radiation chemistry of gases has been studied with α-particles, or more recently, with electron beams. In most practical cases almost all the energy deposited in a gaseous system comes from secondary electrons from the walls of the containing vessel when exposed to γ-radiation.

In contrast to condensed phases, in gases there are no differences between the chemical effects of radiation of different L.E.T. All the primary species produced in what would be isolated spurs, short tracks or blobs in a condensed phase are able to diffuse apart in a gas, and can then react with species from neighbouring sites.

In gases there will be no *cage effects*. Also, when two reactants come together to form an adduct, the excess energy cannot be readily transferred to the nearest neighbours. Therefore, when unstable *transition intermediates* are formed,

$$A + B \longrightarrow AB^* \tag{5.4}$$

they may fly apart, giving new products,

$$AB^* \longrightarrow C + D \tag{5.5}$$

or the original reactants, with no net reaction.

$$AB^* \longrightarrow A + B \tag{5.6}$$

When one of the reactants is an electron, reaction 5.5 is called *dissociative electron capture*. The geometry of the containing vessel can be important because, if the unstable transition intermediate collides with the wall, the excess energy can be transferred to the wall. At higher pressures, the concentration of molecules which can act as a *third body*, to remove excess energy in the formation of the adduct, will be greater, and hence the product yields higher. This 'third body' may be one of the reactants, e.g. B, or it may be another, inert component of the irradiated mixture. Its importance is frequently indicated by writing the reaction in the form,

$$A + B + B \longrightarrow AB + B \tag{5.7}$$

or

$$A + B + M \longrightarrow AB + M \tag{5.8}$$

The third-body efficiency of a component can be evaluated from its effect on product yields.

Mass spectrometry

Gases have another advantage over condensed systems in that considerable information about possible ionic intermediates and ion–molecule reactions occurring in radiolysis can be obtained by comparison with cracking patterns in mass spectrometers. The mass spectrometer, of course, operates by irradiation of a gas with low-energy (about 50 eV = 8 aJ) electrons and measures the amounts of different ions.

Although mass spectrometers normally operate at very low pressures, whereas radiolysis is carried out at much higher pressures, it is probable that such differences in concentration do not affect the primary ions, but only their subsequent reactions. Thus the distribution pattern of the earliest ionic species can be determined. The lifetime of metastable species can also be evaluated.

By using tandem mass spectrometers, others with special high-pressure sources, and molecular beams, the behaviour of charged species in the gas phase has been followed in some detail. The ions may undergo neutralization,

$$M^{\bullet +} + e^- \longrightarrow M \qquad (5.9)$$

$$M^{\bullet +} + B^{\bullet -} \longrightarrow MB \qquad (5.10)$$

or charge transfer,

$$A^+ + B \longrightarrow A + B^+ \qquad (5.11)$$

or ion–molecule reactions.

THE PRIMARY REACTIONS

The primary reactions of ionizing radiation with a molecule in the gas phase may be classified as ionization,

$$M \rightsquigarrow M^{\bullet +} + e^- \qquad (5.12)$$

excitation,

$$M \rightsquigarrow M^* \qquad (5.13)$$

dissociation,

$$M \rightsquigarrow A^\bullet + B^\bullet \qquad (5.14)$$

or dissociative ionization

$$M \rightsquigarrow A^+ + B^\bullet + e^- \qquad (5.15)$$

$$M \rightsquigarrow A^+ + B^- \qquad (5.16)$$

The electrons produced may be captured to form a stable ion, which can exist for a finite time before being neutralized,

$$M + e^- \longrightarrow M^{\bullet -} \qquad (5.17)$$

or undergo dissociative electron capture.

$$M + e^- \longrightarrow A^\bullet + B^- \qquad (5.18)$$

Sometimes the ionic species are formed in an excited state.

$$M + e^- \longrightarrow M^{\bullet -*} \qquad (5.19)$$

Ion–molecule reactions

Ion–molecule reactions—reactions between charged ionic species and stable neutral molecules—are often found in gas-phase radiolysis. Usually, in the case of organic molecules, a proton is transferred. The charge is usually found on the larger of the two product species.

$$CH_4{}^{\cdot +} + CH_4 \longrightarrow CH_5{}^{+} + CH_3{}^{\cdot} \qquad (5.20)$$

If there is excess energy available, the charged product may be in an excited state,

$$CH_4{}^{\cdot +} + CH_4 \longrightarrow CH_5{}^{+*} + CH_3{}^{\cdot} \qquad (5.21)$$

and may dissociate.

$$CH_5{}^{+*} \longrightarrow CH_3{}^{+} + H_2 \qquad (5.22)$$

The ionic product may undergo other ion–molecule reactions.

$$CH_3{}^{+} + CH_4 \longrightarrow C_2H_5{}^{+} + H_2 \qquad (5.23)$$

Ion lifetimes

Ion lifetimes can be measured by applying an intermittent electric field across two plate-electrodes in the gas, which is exposed to continuous

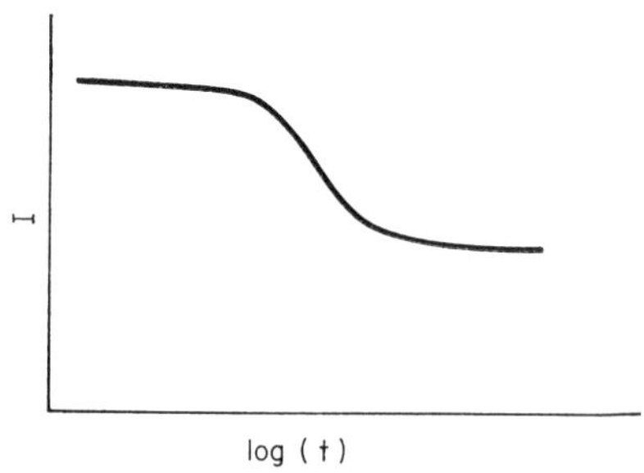

Fig. 5.3 Determination of ion lifetimes by using an intermittent electric field. (t) duration of field-free interval, (I) current.

irradiation. When the electric field is on, all the ions produced are removed, and a current flows between the two electrodes. When the electric field is off, no current flows and the ions which are produced build up and are removed by the normal neutralization reactions in the gas phase, and at the walls of the vessel. As the time interval for which the field is off is progressively decreased, the normal ion-loss processes become less and less effective, since they are normal rate-controlled processes and the current flowing during the periods when the field is on increases

(Fig. 5.3). This increase occurs when the field-free period is of the order of the ion lifetime.

This intermittent-field method resembles in many respects the rotating-sector method used in photochemistry to measure radical lifetimes and the interrupted-dose technique described in Chapter 4.

Excited molecules

Excited molecules can be formed directly by the action of high-energy radiation on a component,

$$M \rightsquigarrow M^* \tag{5.24}$$

or by neutralization of ions.

$$M^+ + e^- \longrightarrow M^* \tag{5.25}$$

$$A^+ + B^- \longrightarrow AB^* \tag{5.26}$$

The excitation may be transferred to another molecule—often called *collisions of the second kind*—perhaps to a minor component,

$$M^* + m \longrightarrow M + m^* \tag{5.27}$$

or may be emitted as a quantum of light,

$$M^* \longrightarrow M + h\nu \tag{5.28}$$

or it may be lost by collision with other molecules or with the walls of the containing vessel and appear as heat. Alternatively, the excited molecules can decompose into radicals,

$$M^* \longrightarrow A\cdot + B\cdot \tag{5.29}$$

or into molecular products, one of which is frequently hydrogen (H_2).

$$M^* \longrightarrow C + D \tag{5.30}$$

Such a unimolecular decomposition is rather slower than energy loss by collision, and hence less likely.

Excited molecules are more reactive than the same molecules in their ground state. Radicals can also be produced in an excited state, when they are known as *hot radicals*. Uncertainties regarding the state of excitation of some of these species is one of the difficulties in interpreting radiation chemistry reactions. Excited species are capable of reactions which are endothermic, and hence improbable for the same species in the ground state. They can also react on the first collision, as they have the energy to overcome the activation barrier.

Triplet states

The *triplet state* is a special type of electronically excited molecule. It is produced when an electron is excited (usually indirectly) to a higher energy state, in which there is already an electron with the same spin. Thus, the triplet state is an excited molecule with two unpaired electrons and may be considered as a *diradical.* Triplet states have relatively long lifetimes, because reversion to the ground state can occur only via a forbidden transition.

Radical reactions

The radicals formed in gas-phase radiolysis are able to undergo the usual types of radical reactions, such as addition to a double bond or aromatic ring,

$$R_1\cdot + R_2C{=}CR_3 \longrightarrow R_1R_2C{-}\dot{C}R_3 \qquad (5.31)$$

abstraction,

$$R\cdot + AB \longrightarrow RA + B\cdot \qquad (5.32)$$

or charge transfer

$$R\cdot + A^- \longrightarrow R^- + A\cdot \qquad (5.33)$$

The radical formed in such reactions is usually less reactive than the attacking radical. However, chain reactions can occur, involving perhaps several thousand reactions, as in polymerization.

Radicals can disappear by a combination reaction,

$$A\cdot + B\cdot \longrightarrow AB \qquad (5.34)$$

or by disproportionation,

$$2R\cdot \longrightarrow C + D \qquad (5.35)$$

or sometimes by reaction with, or adsorption on, the walls of the containing vessel. Reaction with oxygen, which is a diradical, can sometimes give a relatively stable peroxy radical.

$$R\cdot + \cdot O{-}O\cdot \longrightarrow R{-}O{-}O\cdot \qquad (5.36)$$

By these reactions a radical chain reaction may be terminated.

INVESTIGATIONAL TECHNIQUES

Essex technique

In some systems, the products from the decomposition of neutral excited molecules can be distinguished from those from ion–molecule reactions.

If a sufficient electric potential is applied between two electrodes inserted in a gas which is being irradiated, a saturation current flows (cf. Figs. 5.1. and 5.2). Essex showed that the electrons comprising this current and passing through the gas are not sufficiently energetic to cause ionization and increase the number of ion pairs formed. They can, however, cause further excitation. The yields of products from the decomposition of excited molecules will therefore be greater than in the absence of the electric field. On the other hand, the yields of products from ion–molecule reactions will be lower, because the removal of ions by the electric field will prevent some of the ion–molecule reactions from

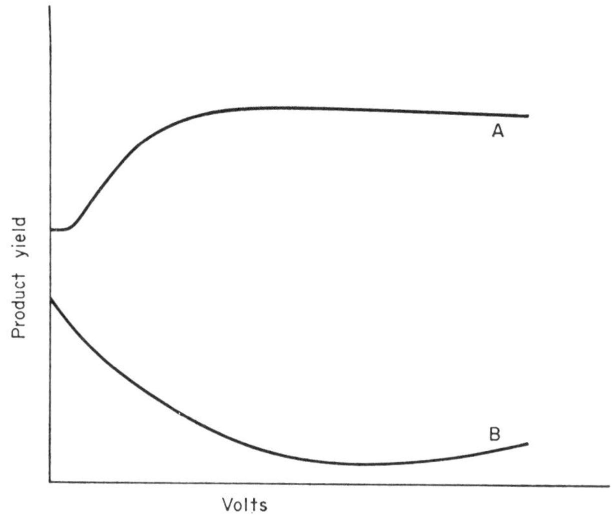

Fig. 5.4 Variation of product yield with increasing applied voltage. (A) product arising from excitation. (B) product arising from ion recombination.

occurring. This method of separating types of reactions is illustrated in Fig. 5.4.

Scavengers

Scavengers can be used to interrupt reactions of specific radicals. From the changes in the final, stable products it is possible to determine the relationships between certain products and their radical precursors. Using the methods of competition kinetics, it is possible to evaluate relative rate constants for these reactions, provided there are no complicating reactions. Nitrous oxide (N_2O) and sulphur hexafluoride (SF_6) are specific scavengers for thermal electrons.

Nitrous oxide undergoes dissociative electron capture with electrons

$$e^- + N_2O \longrightarrow N_2 + O\cdot^- \qquad (5.37)$$

and the N_2O reacted, or the N_2 produced, is a measure of the electron yield. However, it is not a good measure, due to subsequent reactions of $O^{\cdot -}$,

$$O^{\cdot -} + N_2O \longrightarrow N_2 + O_2^{\cdot -} \qquad (5.38)$$

$$O^{\cdot -} + N_2O \longrightarrow NO^{\cdot} + NO^- \qquad (5.39)$$

or reaction with excited species

$$M^* + N_2O \longrightarrow N_2 + O + M \qquad (5.40)$$

Nitric oxide (NO) has been used as a general radical scavenger and ammonia (NH_3) as a scavenger for positive ions. Iodine will act as both an electron trap (to form I^-) and a radical scavenger.

Isotopic labelling

The use of isotopically labelled molecules can often help in sorting out radiolysis mechanisms. If two compounds of different isotopic compositions are mixed, then the resultant isotopic scrambling after irradiation can give a quantitative measure of the recombination reactions which occur.

TYPICAL EXAMPLES

The following systems illustrate the types of reactions which have been proposed to account for the product yields found in irradiated gases. Techniques such as mass spectrometry and radical scavenging and the effects of electric fields, pressure and dose rate on product yields have all contributed to our understanding of the radiation chemistry of the systems.

Radiolysis of oxygen

One of the simplest gases, on which a considerable amount of work has been done, is oxygen. On irradiation, ozone is produced, but the measured yields have varied greatly; being about $G(O_3) = 10$ for γ-rays and 6 for α-particles. The initial value for electron irradiation was found by pulse radiolysis to be 13.8. There is continuing interest in this system, for its own sake, because of the importance of this reaction in the atmosphere and in view of the possible application of radiation industrially in the production of ozone and in the fixation of nitrogen.

There are five primary species produced by the radiation:

$$O_2 \rightsquigarrow O_2^{\cdot +} + e^- \qquad (5.41)$$

$$O_2 \rightsquigarrow O^{\cdot +} + O + e^- \qquad (5.42)$$

$$O_2 \rightsquigarrow 2O \qquad (5.43)$$

$$O_2 \rightsquigarrow O_2^* \qquad (5.44)$$

Neutralization of the ions will produce oxygen atoms

$$O_2{\cdot}^+ + e^- \longrightarrow 2O \quad (5.45)$$

$$O{\cdot}^+ + e^- \longrightarrow O \quad (5.46)$$

Electron attachment to O_2 molecules can occur, forming $O_2{\cdot}^-$ ions

$$O_2 + e^- \longrightarrow O_2{\cdot}^- \quad (5.47)$$

Ozone is believed to be formed mainly by the reaction:

$$O + O_2 + M \longrightarrow O_3 + M \quad (5.48)$$

Either the wall of the containing vessel or a second molecule of oxygen acts as a 'third body' to remove excess energy.

The rate constant for reaction 5.48 has been measured by a number of techniques, mainly using thermal or photochemical generation of oxygen atoms, and is about 2–3 $\times$ $10^8 M^{-2} s^{-1}$.

A serious complication is the back reaction—the radiolysis of ozone—which occurs by a chain reaction. The mechanism is uncertain, but trace impurities are believed to be important. Consequently, useful studies of the radiolysis of oxygen can be made only at low conversions.

Some contribution to the ozone formation may also come from the reaction of excited oxygen molecules,

$$O_2^* + O_2 \longrightarrow O + O_3 \quad (5.49)$$

but there will be a competing deactivation process.

The pressure of the oxygen affects the radiolysis mechanism and the ozone yield, which decreases at high pressures. Electron capture by the parent ion will then have a higher probability. More important, cluster formation can be expected at pressures greater than one atmosphere. Also, at high pressures the reaction

$$2O + O_2 \longrightarrow 2O_2 \quad (5.50)$$

will become significant.

The reactions shown above explain the radiolysis of oxygen, but ion–molecule reactions involving positive or negative ions are also possible.

Studies have also been made on the radiolysis of oxygen–argon mixtures including pulse radiolysis. Argon acts as the 'third body' to remove energy.

$$O + O_2 + Ar \longrightarrow O_3 + Ar \quad (5.51)$$

The rate constant of this reaction can be determined more satisfactorily than that of equation 5.48, since [Ar] can be varied independently of $[O_2]$. It has been shown that the efficiency of Ar as a third body is about 0.6 that of oxygen.

Para–ortho hydrogen conversion

Hydrogen exists in two forms, the ortho form, in which the nuclear spins are parallel, and the para form, in which they are opposed, or anti-parallel.

H–H

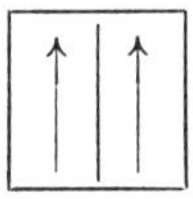

ortho

H–H

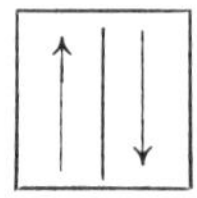

para

The para form is the stable configuration at very low temperatures, but at room temperature the equilibrium mixture contains 75% of the ortho form. However, the rate of interconversion is always very slow in the absence of a suitable catalyst, such as platinum black, or an initiator, such as ultraviolet light or ionizing radiation. This reaction has been studied extensively with thermal, photochemical and radiation initiation, and has been found to proceed by a chain reaction.

The α-particle-initiated reaction was the first accepted case in radiolysis in which excitation, as well as ionization, was demonstrated to give rise to free radicals and chain reactions.

Parent positive ions, electrons and excited molecules are produced directly by the radiation.

$$H_2 \rightsquigarrow H_2{}^{\cdot +} + e^- \tag{5.52}$$

$$H_2 \rightsquigarrow H_2{}^* \tag{5.53}$$

Hydrogen atoms are then produced by decomposition of the excited molecules and $H_3{}^+$ ions by ion–molecule reactions.

$$H_2{}^* \longrightarrow 2H\cdot \tag{5.54}$$

$$H_2{}^{\cdot +} + H_2 \longrightarrow H_3{}^+ + H\cdot \tag{5.55}$$

Two chain reactions for para–ortho conversion are possible—

(i) free-radical

$$H\cdot + p\text{-}H_2 \longrightarrow o\text{-}H_2 + H\cdot \tag{5.56}$$

(ii) ionic

$$H_3{}^+ + p\text{-}H_2 \longrightarrow o\text{-}H_2 + H_3{}^+ \tag{5.57}$$

Experimentally, the kinetic chain length has been found to be about 1000. Neutralization of the $H_3{}^+$ ions will produce more hydrogen atoms.

$$H_3{}^+ + e^- \longrightarrow 3H\cdot \tag{5.58}$$

$$H_3{}^+ + e^- \longrightarrow H_2 + H\cdot \tag{5.59}$$

Removal of hydrogen atoms—and termination of the free radical chain—occurs by dimerization.

$$H\cdot + H\cdot \longrightarrow H_2 \tag{5.60}$$

The free radical chain mechanism in radiolysis is suggested by the analogous thermal and photochemical reactions, and by oxygen inhibition, attributed to

$$H\cdot + O_2 \longrightarrow HO_2\cdot \tag{5.61}$$

However, trace amounts of krypton and xenon markedly reduce the radiolysis yield, but do not affect the photochemical reaction. The effect can be explained by charge transfer from an ionic chain carrier to the rare-gas atom, which has a lower ionization potential. Therefore, it appears that some of the radiolysis chain reactions are free radical and some ionic.

Nitrous oxide

Nitrous oxide is particularly interesting, because the reaction sequence in radiolysis is fairly well understood and because it is the most widely used dosimeter in the gas phase.

After irradiation, the unreacted nitrous oxide, and other oxides of nitrogen, are usually frozen out in a liquid-air trap and the nitrogen produced is measured. $G(N_2) = 10.2$, independent of pressure from 200 to 600 mmHg and many other variables, but does increase significantly at temperatures higher than about 50°C. The W value for nitrous oxide is 30 eV per ion pair. At low doses $G(-N_2O) = 12$, so M/N is about 4. This indicates that decomposition of neutral excited molecules is important, or that some short-chain reactions occur. The initial reactions are ionization and excitation.

$$N_2O \rightsquigarrow N_2O\cdot^+ + e^- \tag{5.62}$$

$$N_2O \rightsquigarrow N_2O^* \tag{5.63}$$

According to mass spectrometric evidence the positive ion can decompose,

$$N_2O\cdot^+ \longrightarrow NO^+ + N\cdot \tag{5.64}$$

but of more importance are neutralization,

$$N_2O\cdot^+ + e^- \longrightarrow N_2O^* \tag{5.65}$$

and dissociative electron attachment.

$$N_2O + e^- \longrightarrow N_2 + O\cdot^- \tag{5.37}$$

The excited molecules decompose,

$$N_2O^* \longrightarrow N\cdot + NO\cdot \qquad (5.66)$$

$$N_2O^* \longrightarrow N_2 + O \qquad (5.67)$$

followed by a variety of reactions of the O and N atoms

$$N\cdot + NO\cdot \longrightarrow N_2 + O \qquad (5.68)$$

$$N\cdot + N\cdot + M \longrightarrow N_2 + M \qquad (5.69)$$

$$O + O + M \longrightarrow O_2 + M \qquad (5.70)$$

$$N_2O + O \longrightarrow O_2 + N_2 \qquad (5.71)$$

Reactions involving negative ions have been postulated

$$O\cdot^- + N_2O \longrightarrow N_2 + O_2\cdot^- \qquad (5.38)$$

$$O\cdot^- + N_2O \longrightarrow NO\cdot + NO^- \qquad (5.39)$$

$$NO^- + N_2O \longrightarrow NO\cdot + N_2 + O\cdot^- \qquad (5.72)$$

The last two reactions are not generally accepted by all investigators. If they do occur they form a chain.

There are still uncertainties in the N_2O system, so that it is not possible to accept all the above reactions as being important. Overall, for gas at 600 torr and 24°C, the reaction can be written

$$12N_2O \xrightarrow[16aJ]{100\,eV} 10.2N_2 + 4.2O_2 + 3.6NO \qquad (5.73)$$

Hydrogen–halogen reactions

Hydrogen will react with a halogen, e.g. bromine or chlorine, on irradiation, to produce the hydrogen halide. Thus, the overall reaction is

$$H_2 + X_2 \longrightarrow 2HX \qquad (5.74)$$

For hydrogen–bromine mixtures, $G(HBr)_{max} \approx 9$, but for hydrogen–chlorine mixtures, $G(HCl) \approx 10^5$, indicating a chain reaction.

Hydrogen–halogen mixtures also react to produce the hydrogen halide with thermal and photochemical initiation. Therefore, a free-radical mechanism is likely for the radiolysis reaction. For bromine, the reaction can be explained by the following sequence:

Firstly, production of hydrogen and bromine atoms by the direct action of the radiation.

$$H_2 \rightsquigarrow 2H\cdot \qquad (5.75)$$

$$Br_2 \rightsquigarrow 2Br\cdot \qquad (5.76)$$

Secondly, reaction of hydrogen atoms with bromine molecules to form hydrogen bromide.

$$H\cdot + Br_2 \longrightarrow Br\cdot + HBr \quad (5.77)$$

Thirdly, removal of bromine atoms by dimerization.

$$2Br\cdot \longrightarrow Br_2 \quad (5.78)$$

The high product yield from hydrogen–chlorine mixtures must result from a chain reaction, and this arises from the greater reactivity of the chlorine atom, which permits the essential chain reaction step—

$$Cl\cdot + H_2 \longrightarrow HCl + H\cdot \quad (5.79)$$

However, this free radical explanation of the overall reaction does not exclude the possibility of ionic, or other, reactions.

Pulse radiolysis

The technique of pulse radiolysis has been used in gases to produce high concentrations of a particular intermediate species and then follow its reactions. In this way rate constants for reaction steps can be obtained. For example, oxygen atoms have been produced by pulse radiolysis of oxygen and then the rate of the reaction—

$$O + O_2 + M \longrightarrow O_3 + M \quad (5.80)$$

measured using a variety of substances as the third body. Similarly pulse radiolysis of hydrogen produces hydrogen atoms, and the rates of their reaction with benzene and toluene in the gas phase have been obtained.

Organic systems

In this chapter our typical examples have been inorganic gases. Because the range of products is more limited, these are somewhat simpler to interpret than are organic gases.

Organic gases undergo similar ionization and excitation processes followed by dissociative, ion–molecule and radical reactions. Because these subsequent reactions are typical of organic compounds, they will be considered in Chapter 7.

FURTHER READING

Ion Molecule Reactions in the Gas Phase, *Adv. Chem. Ser.* No. 58, American Chemical Society (1966).

LAMPE, F. W., FRANKLIN, J. L. and FIELD, F. H., Kinetics of the Reactions of Ions with Molecules in *Progress in Reaction Kinetics*, PORTER, G. (Ed.), Pergamon, Oxford, Vol. 1, p. 67 (1961).

LIND, S. C., HOCHANADEL, C. J. and GHORMLEY, J., *Radiation Chemistry of Gases*, Reinhold, N.Y. (1961).

Chapter 6

Aqueous Solutions

Because of the importance of water and aqueous solutions in chemical processes, in biology and in atomic energy developments, it has been the most intensively studied solvent system. Many authors have said that the main features were firmly established, and been proven wrong. Rather, we will say that the main features are widely accepted and soundly based. Many details remain to be elucidated, but much data has been obtained. It is therefore possible to treat this area in a more systematic fashion than other areas.

Water differs from most other materials in that the immediate radiolytic products do not react readily with the solvent itself. This is because of the high bond strength of the O–H bond [D(H–OH) = 118 kcal/mol (494 kJ/mol)]. There are, however, some reversible interactions, such as hydration. The passage of high-energy ionizing radiation through water or dilute aqueous solutions gives initially electrons, positively charged water ions and excited water molecules.

$$H_2O \rightsquigarrow e^-, H_2O\cdot^+, H_2O^* \qquad (6.1)$$

The electrons lose energy by further collisions until they reach thermal energy (0.025 eV = 0.004 aJ at room temperature) and are then *solvated.* This takes about 10^{-11} seconds.

$$e^- + nH_2O \longrightarrow e^-_{aq} \qquad (6.2)$$

The positive ions, $H_2O\cdot^+$, are thought to be energetically very unstable and to decompose in $\approx 10^{-13}$ seconds, giving H^+ ions and $\cdot OH$ radicals.

$$\left.\begin{array}{ll} & H_2O\cdot^+ \longrightarrow H^+ + \cdot OH \\ \text{or} & H_2O\cdot^+ + H_2O \longrightarrow H_3O^+ + \cdot OH \end{array}\right\} \qquad (6.3)$$

The excited molecules, H_2O^*, have not been identified definitely. They may decompose giving $H\cdot$ atoms and $\cdot OH$ radicals, or perhaps ionize.

$$H_2O^* \longrightarrow H\cdot + \cdot OH \qquad (6.4)$$

$$H_2O^* \longrightarrow e^- + H_2O\cdot^+ \qquad (6.5)$$

Some authors think H_2O^* may exist for microseconds and that this can explain some of the phenomena observed.

Some of the radical products—electrons, hydrogen atoms and hydroxyl radicals—are formed in such proximity, within the spurs, that they can react together. In this way H_2O can be reformed giving no net reaction, or H_2 and H_2O_2 can be formed giving the so-called *molecular products*. The exact precursors of the molecular H_2 and H_2O_2 are still the subject of research. A species $H_3O\cdot$ might also be formed and can perhaps explain some of the results found.

$$H\cdot + \cdot OH \longrightarrow H_2O \quad (6.6)$$

$$e^-_{aq} + \cdot OH \longrightarrow OH^- \quad (6.7)$$

$$e^-_{aq} + e^-_{aq} \longrightarrow H_2 + 2OH^- \quad (6.8)$$

$$H\cdot + H\cdot \longrightarrow H_2 \quad (6.9)$$

$$\cdot OH + \cdot OH \longrightarrow H_2O_2 \quad (6.10)$$

$$e^-_{aq} + H_3O^+ \longrightarrow H_3O\cdot \quad (6.11)$$

So about 10^{-8} seconds after passage of the high-energy ionizing radiation there is

$$H_2O \rightsquigarrow e^-_{aq}, H\cdot, \cdot OH, H_2, H_2O_2, H_3O^+ \text{ and perhaps } H_2O^* \text{ and } H_3O\cdot \quad (6.12)$$

as shown in Fig. 6.1.

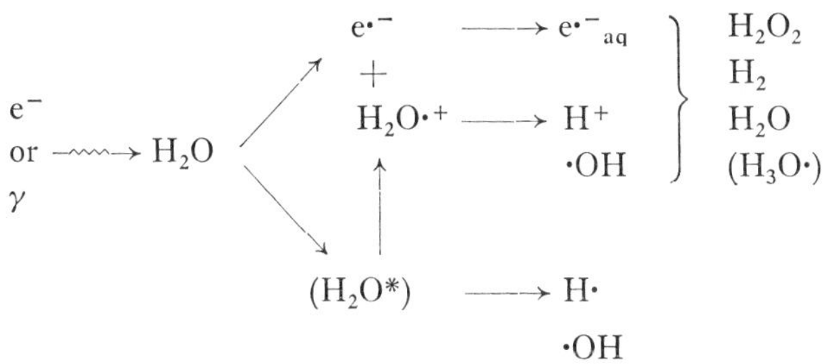

Fig. 6.1 The early processes in the radiolysis of water.

PRIMARY YIELDS

These earliest species which have been detected following the passage of high energy are often called *primary species*. This does not mean that they are the very first entities formed. In Table 6.1 are given the G values for the primary species for light-particle irradiation of acid, neutral and alkaline solutions.

The yields of these species can be determined independently. The accuracy of the measurements must be checked by ensuring that electrical neutrality is preserved – the number of positively charged species must equal the number negatively charged ones. Also there must be a mass balance – the number of hydrogen atoms must be twice the number of oxygen atoms. Alternatively, the number of reducing species (G_e,G_H, $2G_{H_2}$) must equal the number of oxidizing species (G_{OH}, $2G_{H_2O_2}$).

The values given differ from those accepted a few years ago, which showed lower yields for e^- and ·OH in neutral solution. The lower yields may have resulted from insufficient scavenger being present to react with all the primary species. The inhomogeneous distribution of these species means that they should be considered as belonging to two or three groups, the members of which have similar spatial distributions. The second, and then the third, group become available for reaction as the concentration of scavenger is increased progressively. This is better than attempting to explain the results for all scavenger concentrations by using single, average values for the yields of primary species.

The values in alkaline solution are still not accepted by all investigators. It is hard to imagine that the variation with pH is real, because it is unlikely that the decomposition of water molecules will be affected by 1 mM acid or alkali. It is for these and other reasons that some investigators interpret the variation as meaning that another species, which may be activated or excited water, H_2O^*, or the radical $H_3O\cdot$, is formed, with $G \approx 1$.

Table 6.1. G(primary species) in radiolysis of water at low L.E.T.

Species	pH		
	0–2	4–11	13–14
e^-_{aq}	3.65 (e⁻aq + H·)	2.7	3.1
H·		0.55	0.54
·OH	2.95	2.8	2.9
H_2	0.45	0.45	0.45
H_2O_2	0.8	0.7	0.7
H_3O^+		3.6	
OH^-		≈1	
$-H_2O$ (gross)	≈5	≈5	

MOLECULAR PRODUCTS

If the yields of molecular H_2 and H_2O_2 are plotted against the cube root of the concentration of scavengers of e^-_{aq} and ·OH radicals respectively, straight lines result. The one-third power has no particular significance, but can be used empirically to obtain a near estimate of yields.

The hydrogen gas is fairly inert chemically and readily escapes. If the hydrogen is retained in the solution it reacts with the *oxidizing radical* (·OH).

$$H_2 + \cdot OH \longrightarrow H_2O + H\cdot \qquad (6.13)$$

The hydrogen peroxide reacts with the *reducing species* (e^-_{aq} and H·).

$$H_2O_2 + e^-_{aq} \longrightarrow \cdot OH + OH^-$$

$$H_2O_2 + H\cdot \longrightarrow \cdot OH + H_2O \qquad (6.14)$$

So both H_2 and H_2O_2 are destroyed by a chain reaction. In pure water, after a time in a given radiation field, the molecular products are destroyed as fast as they are formed, provided that the hydrogen cannot escape. In other words, a steady state is set up with concentrations of H_2 and H_2O_2 depending on the dose rate. With γ- or electron radiation these concentrations are low and there is very little decomposition. With higher L.E.T. radiation, as in nuclear reactors, the H_2 and H_2O_2 concentrations can rise to higher values.

If the hydrogen is able to escape, oxygen is also produced—ultimately in equivalent amount ($2H_2 + O_2$).

$$H_2O_2 + \cdot OH \longrightarrow H_2O + HO_2\cdot \qquad (6.15)$$

$$HO_2\cdot + HO_2\cdot \longrightarrow H_2O_2 + O_2 \qquad (6.16)$$

$$HO_2\cdot + \cdot OH \longrightarrow H_2O + O_2 \qquad (6.17)$$

Small amounts of impurities can interfere with the chain reactions *6.13* and *6.14*, leading by reactions *6.15* and *6.17* to much greater decomposition.

In contrast with the inert hydrogen gas, the molecular hydrogen peroxide is retained in solution, and in addition to the above reactions, may react with solutes.

In addition to the amounts formed as molecular products, there may also be some hydrogen or hydrogen peroxide formed as stable end products from the reactions of radical species with various solutes.

PRIMARY RADICAL SPECIES

Electrons and hydrogen atoms are powerful reducing agents, while the hydroxyl radical is a powerful oxidizing agent. This means that anything capable of being reduced may be reduced, while anything capable of being oxidized may be oxidized. Sometimes this leads to very little net reaction, but often hydrogen is released from the solution, and oxidizing reactions will then predominate.

Much of the early work was carried out with inorganic solutes in acid solution and could be interpreted in terms of the reactions of two radicals only, H· and ·OH. Later work in neutral solution showed that the reducing radicals were a mixture of *hydrated electrons* and hydrogen atoms and that in acid solutions the electrons were converted into H• atoms.

$$e^{-}_{aq} + H^{+} \rightleftharpoons H\cdot \qquad (6.18)$$

Hydrated electrons and hydrogen atoms act in a similar way stoichiometrically, but not kinetically, and the speeds of their reactions with solutes, such as ferric ion, cupric ion, methanol, acetone and hydrogen peroxide differ considerably. Sometimes, for example with chloroacetic acid solutions, different products result from reactions with the two species.

The hydrated electron

The fact that there was a reducing radical present with unit negative charge was established by showing that, in the presence of inert salts, there was the expected *kinetic salt effect* on the rate of reaction with other charged species.

Shortly afterwards, the *hydrated electron* was identified in pulse radiolysis. In the ensuing years, a great body of data has been built up concerning its reactions. As a fundamental reagent, this will play an important role in chemical theories in many fields, for example those in which electron transfer reactions are important.

It has a broad optical absorption with the maximum at 720 mμ (nm), $\varepsilon = 1.58 \times 10^4\ M^{-1}cm^{-1}$. Its diffusion coefficient is $4.25 \times 10^{-5}\ cm^2/s$, which is not very high, and the reaction radius is about 3 Å (0.3 nm). Its redox potential is about 2.7V.

The hydrated electron may be pictured as being in a *trap* formed by the polarization which it induces in the water-molecule dipoles in its vicinity. In this way the electron can be considered as smeared over its nearest neighbours, four water molecules, and inducing considerably less polarization and orientation in the next outer layer of water molecules.

The absorption at the red end of the visible spectrum means that the solution is an intense blue colour, and is very similar to the solution obtained when an alkali metal such as sodium is dissolved in liquid ammonia. The colour of this particular solution, which has been known for over one hundred years, is also due to solvated electrons, liberated and stabilized as Na^{+} is formed. Such a solution in liquid ammonia or an amine will remain stable for some time.

On the other hand, in water, hydrated electrons react with each other giving hydrogen, or very slowly with pure water, or often very rapidly with impurities or solutes, or with protons in acid solutions. Rate constants (*k*) are shown in units of $M^{-1}\ s^{-1}$ ($dm^3\ mol^{-1}\ s^{-1}$). [$1\ M^{-1}\ s^{-1} = 10^3\ cm^3\ mol^{-1}\ s^{-1} = 10^{-3}\ m^3\ mol^{-1}\ s^{-1} \approx 1\ kg\ mol^{-1}\ s^{-1}$ for dilute aqueous solutions.]

$$e^-_{aq} + e^-_{aq} \longrightarrow H_2 + 2OH^- \qquad k = 5 \times 10^9 \qquad (6.8)$$

$$e^-_{aq} + H_2O \longrightarrow H\cdot + OH^- \qquad k = 16 \qquad (6.19)$$

$$e^-_{aq} + S \longrightarrow S\cdot^- \qquad k = 10^5 - 10^{11} \qquad (6.20)$$

$$e^-_{aq} + H^+ \rightleftharpoons H\cdot \qquad k = 2.3 \times 10^{10} \qquad (6.18)$$

Reaction *6.18* shows that H• atom is an acid and e^-_{aq} is its *conjugate base.* The pKa is about 9.7. This means that, at equilibrium, there are equal amounts of e^-_{aq} and H• at pH 9.7 and a greater proportion of H• at lower pH. Because there is rarely time for equilibrium to be established, reaction *6.18* is usually unimportant above pH 4, but has been observed to occur in neutral solutions of very pure water. In most radiation studies, e^-_{aq} is the only reducing species important above pH 13, due to the reaction,

$$H\cdot + OH^- \longrightarrow e^-_{aq} \qquad k = 2 \times 10^7 \qquad (6.21)$$

Both e^-_{aq} and H• are present in the range of pH 4–11, and H• is the important reducing species below pH 2. Reaction with dihydrogen phosphate ions also gives H• atoms, but the rate is slow, except at high phosphate concentrations.

$$e^-_{aq} + H_2PO_4^- \longrightarrow H\cdot + HPO_4^{2-} \qquad k = 10^7 \qquad (6.22)$$

Hydrated electron reactions with solutes are predominantly electron attachment reactions, i.e. reductions. Therefore, hydrated electrons will react with a substance having a vacant electron orbital, provided that the energy balance favours the addition of an electron.

They react at a diffusion controlled rate, i.e. at every encounter, with almost all free radicals. Aqueous, radiolytic, transient species are almost always free radicals.

Methanol and ethanol are often used to scavenge hydrogen atoms and hydroxyl radicals.

$$CH_3OH + H\cdot, \cdot OH \longrightarrow \cdot CH_2OH + H_2, H_2O \qquad (6.23)$$

$\dot{C}H_2OH$ and $CH_3\dot{C}HOH$ are radicals which do not react readily with hydrated electrons.

Molecular oxygen is a *diradical* and has two unpaired electrons. It reacts readily.

$$\cdot O_2\cdot + e^-_{aq} \longrightarrow O_2\cdot^- \qquad k = 1.9 \times 10^{10} \qquad (6.24)$$

The product is important in aerated solution and will be considered separately.

The reaction with nitrous oxide is important, because it is specific for electrons, and also because they may be converted by this means into oxidizing radicals (·OH) via the unstable adduct $N_2O\cdot^-$.

$$
\begin{array}{ccccc}
N_2O + e^-_{aq} & \longrightarrow & [N_2O\cdot^-] & \longrightarrow & N_2 + O\cdot^- \\
 & & + H^+ & & + H^+ \\
 & & \downarrow & & \downarrow \\
 & & N_2\dot{O}H & \longrightarrow & N_2 + \cdot OH
\end{array}
\qquad (6.25)
$$

Hydrated electrons react with many uncharged molecules and monovalent cations (rate constants of about 2–3 × 10^{10} M^{-1} s^{-1}), and monovalent anions (rate constants of about 1–2 × 10^{10} $M^{-1}s^{-1}$) at every encounter. However, most of the reaction rates of hydrated electrons are lower than these values (see Fig. 6.2), and this implies some steric or energetic restriction on the reaction. Reaction *6.18* with protons, usually designated in aqueous solution as H_3O^+, is somewhat slower than diffusion controlled.

Hydrated electrons do not react readily with alkali metal, alkali earth, ammonium or ferrocyanide ions or with some EDTA complexes. They react readily with most other metal ions or metal complexes, giving an ion of lower charge or zero charge, which in some cases reacts subsequently with water, liberating hydrogen. Metal complexes may have reactivities different from their metal ions. The ligand field may change the electron density. Other ligands, such as the halides, are thought to channel electrons into the metal atom.

Hydrated electrons react readily with strong oxidizing agents, many sulphur-containing compounds, halogen-organic or nitro-compounds, nitrite and nitrate ions, but not with many fluoro compounds.

The electron is the simplest *nucleophilic* reagent. Generally, hydrated electrons will add on to organic compounds where there is a residual or partial positive charge. They react with the π-electrons of many olefinic compounds, but not readily with ethylene. The reactivity is enhanced by conjugation of double bonds.

Compounds containing carbonyl groups are generally reactive. Those ketones and aldehydes which are not hydrated are reactive. An electron-donating group, such as hydroxy, alkoxy or amino, reduces the reactivity. This means that acids, particularly polybasic acids, carboxylates and esters are not reactive. Formic and acetic acids in the acid form do, however, react. Alcohols and phenols do not.

In reaction with aromatic compounds, e^-_{aq} acts as a very reactive nucleophilic reagent. Its reactivity depends on the π-electron density in the aromatic ring and correlates well with the Hammett σ-function. In the case of halogen-substituted aromatics the reactivity with the halogen atom is superimposed on the reactivity with the ring. In heterocyclic

(1 $M^{-1} s^{-1}$ = 1 $dm^3 mol^{-1} s^{-1}$ ≈ 1 $kg mol^{-1} s^{-1}$)

$M^{-1} s^{-1}$										$M^{-1} s^{-1}$
10^{11}	$Co(NH_3)_6^{3+}$									10^{11}
		Ag^+			NO	CS_2	CCl_4	$PhNO_2$		
	H^+		CrO_4^{2-}	O_2	Acrylamide					
			IO_4^-				Purine	Pyridine	Thymine	
10^{10}		NO_3^-			H_2O_2			PhI	Cystine ±	10^{10}
			NO_2^-	ClO^-	N_2O	CO_2	Fumarate Acetone	Acetaldehyde	Naphthalene	
			$Fe(CN)_6^{3-}$				Thiourea	$PhCOO^-$		
							Chloroacetate			
	Zn^{2+}									
10^9					CO					10^9
								PhCl		
	Fe^{2+}						Acetic acid Formic acid		Peptides	
10^8										10^8
								$PhNH_2$		
									Proline	
10^7			$H_2PO_4^-$					PhH		10^7
									Glycine Alanine	
LOW	NH_4^+		HCO_3^-	ClO_4^-	H_2	CH_4	HCHO $HCOO^-$	PhOH		LOW
	$Zn(EDTA)^{2-}$		CN^-	Cl^-		C_2H_4	CH_3COO^-			
	Na^+		$Fe(CN)_6^{4-}$			EtOH	MeOH Citrate		Sugars	
	Ca^{2+}		SO_3^{2-}	SO_4^{2-}		EDTA	Ethers Esters	CH_3NH_2		

Fig. 6.2 Rate constants for the reactions of hydrated electrons with various solutes.

rings a tertiary nitrogen, =N–, increases the reactivity, whereas an –NH– group decreases the reactivity.

The most reactive grouping determines the reactivity of compounds containing many groups.

The reactions of hydrated electrons with many biochemical compounds have been investigated, and an outline of this work is given in Chapter 9.

The hydrogen atom

Rather less is known about the reactions of the *hydrogen atom*. It reduces most substances that can be easily reduced, but is not as strong a reducing agent as the hydrated electron. It can also act by abstracting another atom of hydrogen from a compound. In acid solutions, hydrated electrons are converted to hydrogen atoms by reaction with protons (reaction *6.18*), whereas in alkaline solution hydrogen atoms react with hydroxyl ions giving hydrated electrons (reaction *6.21*).

The rates of reaction of hydrogen atoms with a number of solutes in aqueous solution have been measured, mainly by competition methods, and a selection is given in Fig. 6.3.

Many of the reactions reported in the literature prior to 1960, in neutral or alkaline solution, were actually those of the hydrated electron, and the data require a new interpretation. In acid solutions, the hydrated electrons are generally converted to hydrogen atoms, so the early interpretations were correct.

One important reaction is that with oxygen, giving the hydroperoxyl radical, $HO_2\cdot$, the acid form of the product, $O_2\cdot^-$, of electron–oxygen reaction *6.24*. We will deal separately with the hydroperoxyl radical.

$$\cdot O_2\cdot + H\cdot \longrightarrow HO_2\cdot \qquad k = 2 \times 10^{10} \qquad (6.26)$$

The main reactions with organic compounds are hydrogen abstraction and double bond or ring addition. This means that a wide range of compounds react, including hydrocarbons. PhH is benzene.

$$RH + H\cdot \longrightarrow H_2 + R\cdot \qquad (6.27)$$

$$PhH + H\cdot \longrightarrow H\dot{P}hH \qquad (6.28)$$

Hydrogen abstraction occurs more readily from a –CH or $-CH_2$ group than from a $-CH_3$ group, and hydrogen atoms α to a –COOH or –OH group appear to be favoured.

$$CH_3OH + H\cdot \longrightarrow H_2 + \dot{C}H_2OH \qquad k = 2 \times 10^6 \qquad (6.29)$$

$$CH_3CH_2OH + H\cdot \longrightarrow H_2 + CH_3\dot{C}HOH \qquad k = 2 \times 10^7 \qquad (6.30)$$

H atoms can abstract halogen atoms from certain organic halogen

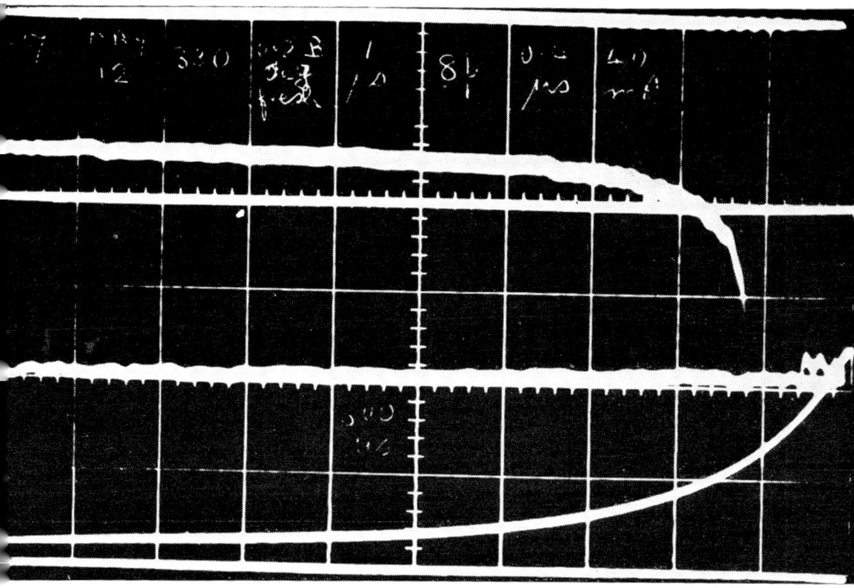

Plate 5. Pulse radiolysis trace. The upper trace spans 10 microseconds and shows the growth (right to left) of an absorbing species. The lower trace shows its subsequent decay (right to left) and spans 5 milliseconds (p. 54).

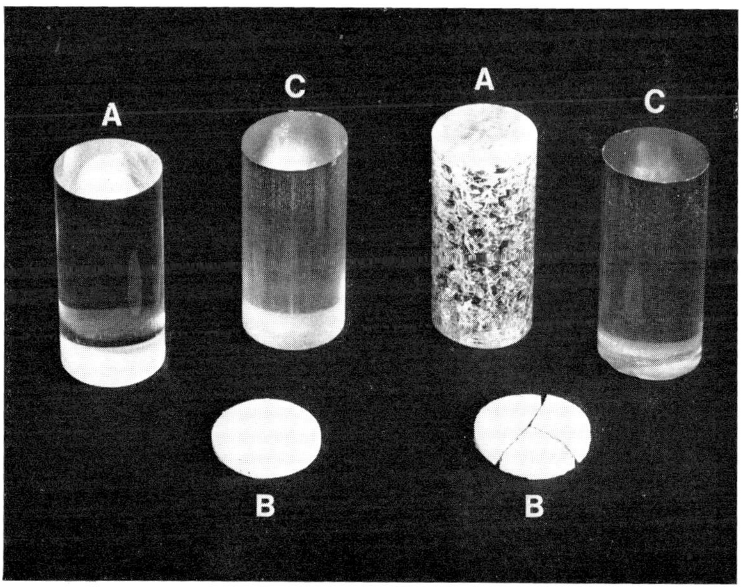

Plate 6 Degradation of polymers by 60 Mrad of gamma radiation. **A** poly(methyl methacrylate), **B** poly(tetrafluoroethylene), **C** poly(styrene). Specimens on the left are unirradiated controls (p. 121).

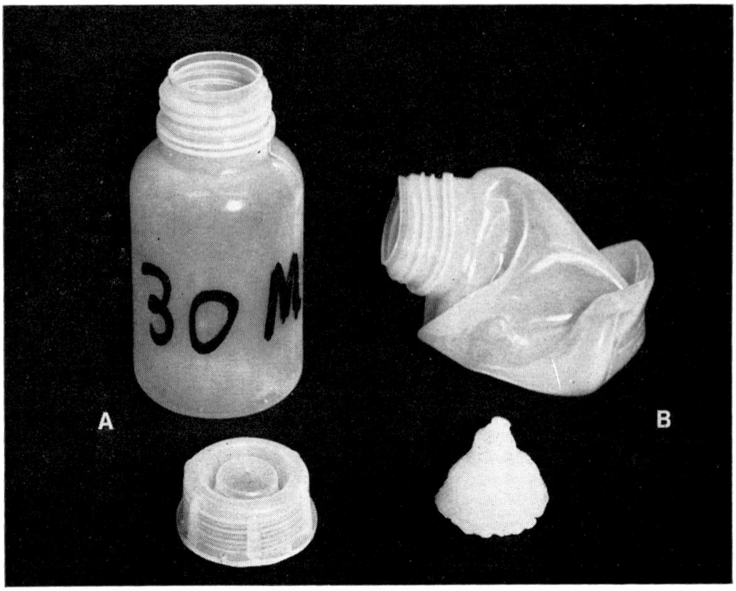

Plate 7 Heat resistance of poly(ethylene) after crosslinking by irradiation. Result of 5 minutes in oven at 120°C. **A** 30 Mrad of gamma radiation, **B** unirradiated (p. 121).

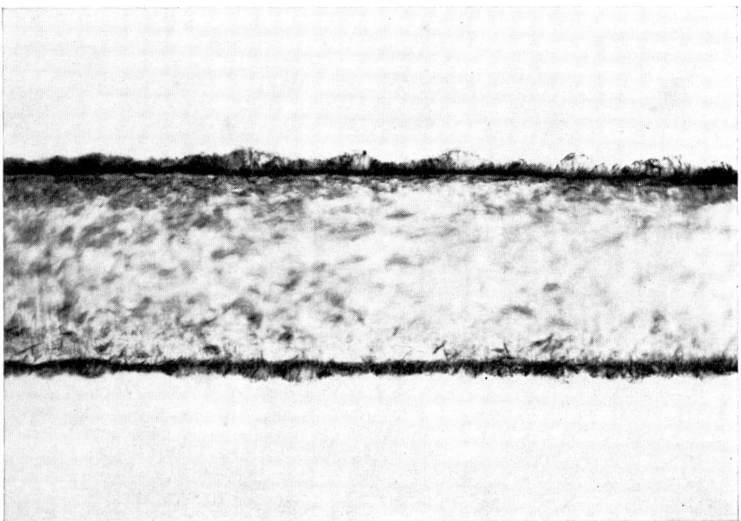

Plate 8 Micrograph of graft copolymer of styrene on poly(tetrafluoroethylene). The grafted poly(styrene) layer on either side of poly(tetrafluoroethylene) sheet is shown in cross-section. [Sample by courtesy of Dr. H. A. J. Battaerd. Microphotograph by Mr. J. Wyatt] (p. 125).

(1 $M^{-1} s^{-1}$ = 1$dm^3 mol^{-1} s^{-1}$ ≈ 1 $kg mol^{-1} s^{-1}$)

$M^{-1} s^{-1}$									$M^{-1} s^{-1}$
10^{11}									10^{11}
				I_2					
10^{10}	Ag^+	CrO_4^{2-}		O_2 Br_2					10^{10}
			$Fe(CN)_6^{3-}$				PhOH	Cystine Uracil	
		CN^-							
10^9		NO_2^-					$PhNO_2$ $PhCOO^-$		10^9
					$HCOO^-$		$PhNH_2$		
10^8	Fe^{3+}			H_2O_2				Pyridine Sugars	10^8
		OH^-							
	Fe^{2+}				EtOH			Glycine	
10^7		NO_3^-							10^7
					MeOH	Chloroacetate Acetone Acetic acid			
LOW	$Co(NH_3)_6^{3+}$ H^+ Zn^{2+}		HCO_3^- ClO_4^-	CO_2 N_2O				Esters	LOW

Fig. 6.3 Rate constants for the reactions of hydrogen atoms with various solutes.

compounds. This often occurs in competition with hydrogen abstraction from the same compound.

$$ClCH_2COO^- + H\cdot \longrightarrow H_2 + Cl\dot{C}HCOO^- \quad k = 2.6 \times 10^6 \quad (6.31)$$

$$ClCH_2COO^- + H\cdot \longrightarrow HCl + \dot{C}H_2COO^- \quad k = 2.6 \times 10^5 \quad (6.32)$$

They add readily to the ring in aromatic compounds, such as nitrobenzene, aniline and benzoate ion, in much the same way as an *electrophilic* reagent.

The hydroxyl radical

There is good evidence for claiming that the oxidizing species found in the radiolysis of water is the *hydroxyl radical.* It has been shown that it is uncharged.

The hydroxyl radical has a weak optical absorption. The extinction coefficient, ε, is 370 $M^{-1}cm^{-1}$ at 260 mμ (nm), and increases towards the far-ultraviolet. This can be used only with difficulty to measure rates of reaction with solutes by following the disappearance of the radical. In some cases the immediate product is well defined and the absolute rate of reaction can be measured by following its appearance by pulse radiolysis techniques. Most of the known rates of reaction have been relative rates determined by competition methods (Fig. 6.4). Values given are the 'best' available. There is some lack of agreement.

In alkaline solution $\cdot OH$ radical reacts with OH^-. The pK is 11.9,

$$OH^- + \cdot OH \rightleftharpoons H_2O + O\cdot^- \quad k = 3.6 \times 10^8 \quad (6.33)$$

and the conjugate base can react with oxygen in aerated solution.

$$O\cdot^- + O_2 \longrightarrow O_3\cdot^- \quad k = 3 \times 10^9 \quad (6.34)$$

In the absence of solutes which react with them, $\cdot OH$ radicals react with reducing radicals (reactions *6.6* and *6.7*) or their adducts, re-forming water, or with each other, giving hydrogen peroxide.

$$\cdot OH + \cdot OH \longrightarrow H_2O_2 \quad k = 5 \times 10^9 \quad (6.35)$$

They are powerful oxidizing agents and undergo four types of reaction. charge transfer,

$$X^- + \cdot OH \longrightarrow X\cdot + OH^- \quad (6.36)$$

hydrogen abstraction,

$$RH + \cdot OH \longrightarrow R\cdot + H_2O \quad (6.37)$$

addition,

$$RH + \cdot OH \longrightarrow H\dot{R}OH \quad (6.38)$$

and displacement.

$$RI + \cdot OH \longrightarrow ROH + I\cdot \quad (6.39)$$

$(1\ M^{-1}\ s^{-1} = 1\ dm^3\ mol^{-1}\ s^{-1} \approx 1\ kg\ mol^{-1}\ s^{-1})$

$M^{-1}\ s^{-1}$										$M^{-1}\ s^{-1}$
10^{11}										10^{11}
	CNS^-				PhOH					
10^{10}	I^-	$Fe(CN)_6^{4-}$	NO							10^{10}
	SO_3^{2-}	NO_2^-	EDTA	{Acrylamide	PhH	$PhCOO^-$	{$PhNO_2$	PhCl	Thymine	
	Br^-	Cl^-	acid	{Acrylamide		PhCOOH	{$PhNO_2$			
				Formate Ether						
				EtOH	Pyridine				Glucose	
10^9	Br^- neut									10^9
	Fe^{2+}			MeOH					DNA	
	Cu^{2+}	CO_3^{2-}	OH^-	Esters					Peptides	
10^8				Acetone						10^8
				Acetate						
			H_2							
			H_2O_2							
10^7	HCO_3^-								Glycine	10^7
LOW	Cl^- alk		$H_2PO_4^-$	CO_2 $CHCl_3$						LOW
	CN^-	ClO_4^-	NO_3^-							

Fig. 6.4 Rate constants for the reactions of hydroxyl radicals with various solutes. Some have been measured directly. Others are relative measurements. Bracketted values show independent measurements which differ significantly.

Electron transfer (one electron oxidation) takes place with a variety of anions (halides and pseudo-halides) and with certain metal ions and complexes. The bromide and chloride cases are interesting ones as the rate constant is greater at low pH than in neutral or alkaline solution. The halide or pseudo-halide radical often adds on to one of its parent ions.

$$CNS^- + \cdot OH \longrightarrow \cdot CNS + OH^- \quad k = 2 \times 10^{10} \qquad (6.40)$$

$$CNS^- + \cdot CNS \longrightarrow (CNS)_2\cdot^- \qquad (6.41)$$

Hydrogen abstraction by hydroxyl radicals parallels that by hydrogen atoms, except that it occurs with rather higher rate constants (cf. reactions *6.29* and *6.30*).

$$CH_3OH + \cdot OH \longrightarrow \cdot CH_2OH + H_2O \quad k = 8 \times 10^8 \quad (6.42)$$

$$C_2H_5OH + \cdot OH \longrightarrow CH_3\dot{C}HOH + H_2O \quad k = 1.8 \times 10^9 \quad (6.43)$$

Addition reactions predominate when π-electrons are present, e.g. in ethylenic or aromatic compounds. These reactions have been more intensively investigated than have the addition reactions of hydrogen atoms. The isomeric distribution of final, stable, hydroxy products following ·OH radical addition to substituted benzenes shows that the ·OH radical has some electrophilic character superimposed on the high reactivity expected from a free radical.

Hydroperoxyl radical

The *hydroperoxyl radical*, $HO_2\cdot$, has been identified as a 'primary' species in the tracks of heavy particles in deaerated water. It probably arises from the reaction:

$$\cdot OH + H_2O_2 \longrightarrow H_2O + HO_2\cdot \quad k = 3 \times 10^7 \qquad (6.44)$$

With light particle irradiation, $HO_2\cdot$, or its ionized form $O_2\cdot^-$, is formed in aerated solution by the reaction of hydrated electrons or hydrogen atoms with dissolved oxygen (reactions *6.24* and *6.26*). Oxygen dissolves in water or dilute aqueous solutions to give a solution about 0.26 mM when they are exposed to the air at room temperature. The rate constants for e^-_{aq} and H· reactions with oxygen are 1.9×10^{10} and 2×10^{10} M^{-1} s^{-1} respectively. This means that either a high concentration of solute, or a fairly reactive one, must be present to prevent formation of $O_2\cdot^-$ or $HO_2\cdot$.

$HO_2\cdot$ is an acid with a pK of 4.4, which makes it about as strong as most organic acids. It is unreactive enough with most solutes (except nitromethane) to enable equilibrium to be established, so that $HO_2\cdot$ is formed in acid solution, whereas in neutral solution, $O_2\cdot^-$ is the form found.

$$HO_2\cdot \rightleftharpoons H^+ + O_2\cdot^- \qquad (6.45)$$

Not much is known about the reactions of the species. Usually it disappears by reaction with itself, giving hydrogen peroxide.

$$HO_2\cdot + HO_2\cdot \longrightarrow H_2O_2 + O_2 \quad k = 2 \times 10^6 \tag{6.46}$$

Organic solute radicals

Whereas a great deal is now known about the reactions of the primary radical species with various solutes, rather less is known about the behaviour of the radicals formed. Generally, electron adducts protonate readily, usually by abstracting a hydrogen ion from water, to give a species which may, or may not, be identical with the hydrogen adduct.

$$A + e^-_{aq} \longrightarrow A\cdot^- \tag{6.47}$$

$$A\cdot^- + H_2O \longrightarrow AH + OH^- \tag{6.48}$$

This is one reason why the hydrated electron took so long to discover; the results of its reactions are so nearly the same as those expected for hydrogen atoms.

When a solute of greater electron affinity is present, even in much lower concentration, the electron can be transferred to the second solute, thus diverting the normal reaction sequence and giving a different product.

$$A\cdot^- + B \longrightarrow A + B\cdot^- \tag{6.49}$$

This will give a wrong answer when competition methods are used to measure relative rates of reaction. Also, certain impurities can, in this way, have a profound effect on radiolytic reactions.

In deaerated solutions there are almost equal numbers of oxidizing radical adducts and reducing radical adducts. Recombination reactions giving the original solute and solvent back again,

$$HR\cdot^- + H\dot{R}OH \longrightarrow 2HR + OH^- \tag{6.50}$$

appear to be more favoured than are dimerization or disproportionation reactions between like radicals.

$$2H\dot{R}H \longrightarrow H_2RRH_2 \tag{6.51}$$

$$2H\dot{R}H \longrightarrow HRH_2 + RH \tag{6.52}$$

Radicals formed by hydrogen abstraction may also dimerize

$$2R\cdot \longrightarrow R_2 \tag{6.53}$$

In aerated solutions molecular oxygen adds on to many radicals, and this prevents both recombination and dimerization reactions. Generally speaking, there is more radiolytic reaction found in aerated systems than in deaerated. Oxygen is recognized as a sensitizer in radiation biology.

In addition to preventing recombination reactions, the formation of peroxides is also important. These sometimes eliminate $HO_2\cdot$ or $O_2\cdot^-$, re-forming the solute or giving an hydroxy compound.

$$HR\cdot^-, H\dot{R}H, H\dot{R}OH + O_2 \longrightarrow HRO_2\cdot^-, HRHO_2\cdot, HROHO_2\cdot \quad (6.54)$$

$$HRO_2\cdot^- \longrightarrow O_2\cdot^- + HR$$

$$HRHO_2\cdot \longrightarrow HO_2\cdot + HR \quad (6.55)$$

$$HROHO_2\cdot \longrightarrow HO_2\cdot + ROH \quad (6.56)$$

The most readily observed reactions are therefore oxidizing radical reactions, until all the oxygen is used up. This corresponds to a dose of about 50 krad and means that $G(-O_2)$ is about 5. Thereafter, the oxidizing and reducing reactions will be more or less equally effective, depending upon the chemistry of the solute.

Radicals as acids and bases

Many radicals act as acids or bases. In either form they still have an unpaired electron, but their reaction rates can be different. Some acid–base pairs are listed in Table 6.2. Sometimes other reactions occur so rapidly that equilibrium is not established.

Table 6.2. Transient radicals that are acids.

Acidic Form	pK	Basic Form
$H\cdot$	9.7	e^-
$\cdot OH$	11.9	$O\cdot^-$
$HO_2\cdot$	4.5	$O_2\cdot^-$
$\dot{C}H_2OH$	10.7	$\dot{C}H_2O^-$
$Ph\dot{N}O_2H$	3.2	$PhNO_2\cdot^-$

CONCENTRATED SOLUTIONS

Most of the work on aqueous radiation chemistry has been carried out on dilute solutions. Indirect effects predominate and the chemistry of the primary species from the radiolysis of water is what is being investigated. Slightly higher product yields are obtained above 1 mM, since radicals are also scavenged in the spurs.

In solutions above about 0.1 M, direct effects—the action of radiation on the solute itself—start to appear, although they do not become important until the concentration is greater than about 1 M. Water itself is about 56 M and interaction with radiation is a function of the number of electrons present.

In strong salt solutions the water structure is changed somewhat and so is the environment in which a hydrated electron finds itself. This may be responsible for some of the differences between neutral and strongly alkaline solutions. In strongly acidic solutions, these effects are overshadowed by the conversion of hydrated electrons to hydrogen atoms.

Since, in many practical cases of radiation effects, one is dealing with reasonably concentrated solutions, it is most important to understand the reactions occurring in these solutions. Insufficient cases have been studied thoroughly enough for a pattern to emerge yet.

Protective agents

Small quantities of certain substances will protect some biological systems against radiation damage. These chemicals, thiourea, cysteine, etc., are good scavengers for both oxidizing and reducing radicals and, in particular, can compete with oxygen for the hydrated electron.

TYPICAL EXAMPLES

The following systems have been chosen as typical of the range of reactions found in many aqueous solutions. In many cases, chemical equations can be set up to explain the products found. The proposed reaction scheme can be tested by altering concentrations of solutes and other conditions and by relating product yields quantitatively to the yields of primary species.

Ferrous ion in acid solution

In dilute, aerated, sulphuric acid solutions of ferrous ion the following reactions occur:

$$Fe^{2+} + \cdot OH \longrightarrow Fe^{3+} + OH^- \qquad (6.57)$$

$$H\cdot + O_2 \longrightarrow HO_2\cdot \qquad (6.26)$$

$$Fe^{2+} + HO_2\cdot \longrightarrow Fe^{3+} + HO_2^- \qquad (6.58)$$

$$HO_2^- + H^+ \longrightarrow H_2O_2 \qquad (6.59)$$

$$2Fe^{2+} + H_2O_2 \longrightarrow 2Fe^{3+} + 2OH^- \qquad (6.60)$$

For every 100 eV absorbed, there are G_{OH} molecules of Fe^{3+} formed by reaction *6.57* and G_H by reaction *6.26*, followed by *6.58*. There are two molecules of Fe^{3+} formed by reaction *6.60* from each molecule of H_2O_2. H_2O_2 is formed with yield G_H by the reaction sequence *6.26*, *6.58* and *6.59*. It is also a primary species with yield $G_{H_2O_2}$. We can sum all sources of Fe^{3+}.

$$G(Fe^{3+}) = G_{OH} + G_H + 2G_H + 2G_{H_2O_2}$$

From Table 6.1 $= 2.95 + 3 \times 3.65 + 2 \times 0.8$
$= 15.5$

Thus the reaction scheme and the listed yields of primary species are in agreement with the experimentally determined product yield.

An organic impurity would react to give an increased yield of ferric ion.

$$RH + \cdot OH \longrightarrow R\cdot + H_2O \qquad (6.37)$$

$$R\cdot + O_2 \longrightarrow RO_2\cdot \qquad (6.61)$$

$$3Fe^{2+} + RO_2\cdot + 3H^+ \longrightarrow 3Fe^{3+} + ROH + H_2O \qquad (6.62)$$

But if chloride ion is added, as in the Fricke dosimeter, the normal yield is obtained.

$$Cl^- + \cdot OH + H^+ \longrightarrow Cl\cdot + H_2O \qquad (6.63)$$

$$Fe^{2+} + Cl\cdot \longrightarrow Fe^{3+} + Cl^- \qquad (6.64)$$

In deaerated solution reactions *6.26* and *6.59* are replaced by,

$$Fe^{2+} + H\cdot + H^+ \longrightarrow Fe^{3+} + H_2 \qquad (6.65)$$

$$G(Fe^{3+}) = G_{OH} + G_H + 2G_{H_2O_2} = 8.2$$

Ethyl alcohol

Neutral, aerated, aqueous solutions of ethanol give acetaldehyde.

$$CH_3CH_2OH + \cdot OH \longrightarrow CH_3\dot{C}HOH + H_2O \qquad (6.43)$$

$$CH_3\dot{C}HOH + O_2 \longrightarrow CH_3CHO + HO_2\cdot \qquad (6.66)$$

G(acetaldehyde) $= 2.2$ and should be equal to G_{OH} from reactions *6.43* and *6.66*. $G(H_2O_2) = 3.2$ and should be equal to $G_{H_2O_2} + \frac{1}{2}(G_H + G_e + G_{OH})$ from the molecular H_2O_2 yield and reaction *6.46* following the production of $HO_2\cdot$ or $O_2\cdot^-$ by reactions *6.26*, *6.24*, *6.43* and *6.66*. The following values for yields of primary species can be deduced: $G_{H_2O_2} = 0.7$, $G_H = 0.6$. $G_e = 2.2$ and $G_{OH} = 2.2$. The yields for G_e and G_{OH} are much lower than those deduced from studies of other systems and given in Table 6.1. The reason for this difference is not known. It may be due to an effect of the spatial distribution of ionizing events, additional transient species or some other chemical reactions which occur in the alcohol solution. Alternatively, at the doses used in such studies the products may scavenger some of the primary radical species.

In deaerated solution $CH_3\dot{C}HOH$ is also formed from reaction of the H· atom with ethanol. Two of these radicals dimerize giving a glycol.

$$2CH_3\dot{C}HOH \longrightarrow \begin{matrix} CH_3CHOH \\ | \\ CH_3CHOH \end{matrix} \qquad (6.67)$$

G(glycol) $= 1.6 \approx \frac{1}{2}(G_{OH} + G_H)$.

But in acid, deaerated solution acetaldehyde is also formed.

$$G(\text{glycol}) + G(\text{aldehyde}) = 1.65 + 1.90 \approx \tfrac{1}{2}(G_{OH} + G_H + G_e)$$

$$G(H_2) = 4.2 \approx G_{H_2} + G_H + G_e$$

Benzene

Water will dissolve benzene to give a solution up to 32 mM in concentration. Care must be taken because benzene is volatile from solution. In neutral, deaerated, aqueous solutions of benzene a hydroxycyclohexadienyl radical is formed.

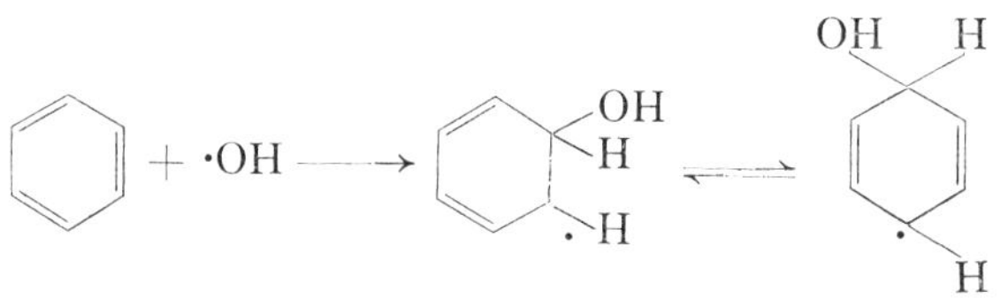

(6.68)

This disappears by second-order reactions,

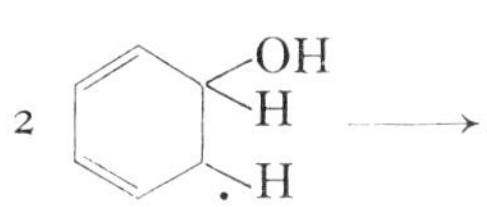

dimer
or $C_6H_5OH + C_6H_7OH$
or $C_6H_5OH + C_6H_6 + H_2O$
phenol

(6.69)

or by reaction with the hydrogen atom adduct.

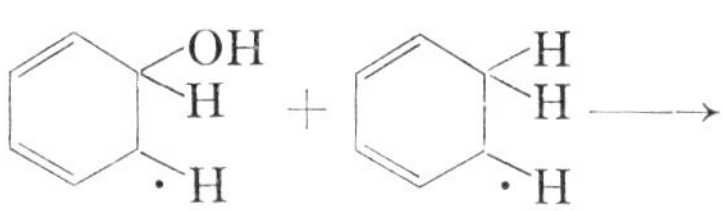

$2C_6H_6 + H_2O$
or $C_6H_5\text{–}C_6H_5 + H_2 + H_2O$
diphenyl

(6.70)

The two hydrogen adducts combine similarly to give dicyclohexadiene or diphenyl and hydrogen.

G(dimer) = 0.2, G(diphenyl) = 0.3, G(phenol) = 0.23. The compound C_6H_7OH has not been definitely identified.

In aerated, neutral solutions a hydroxycyclohexadienyl peroxy radical is formed.

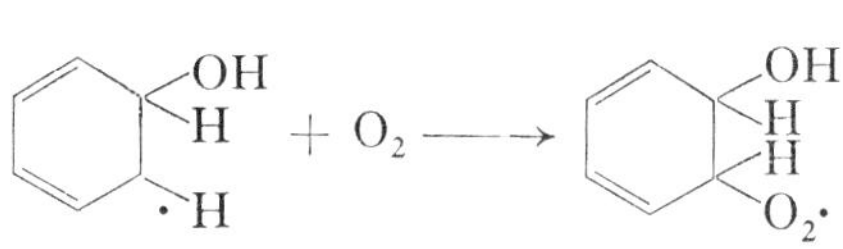

(6.71)

The following reactions then occur:

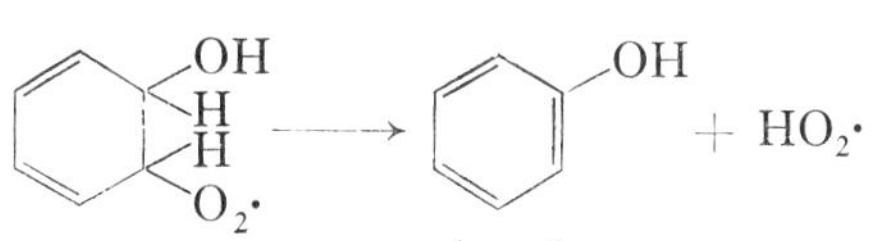

phenol

(6.72)

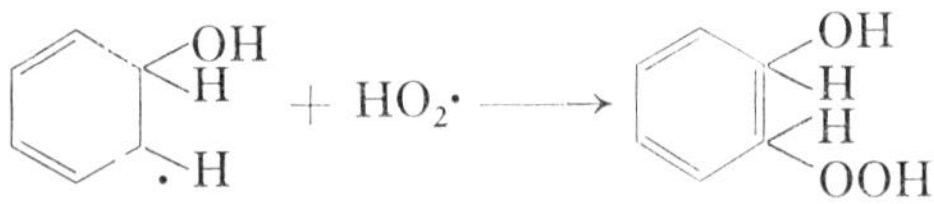

(6.73)

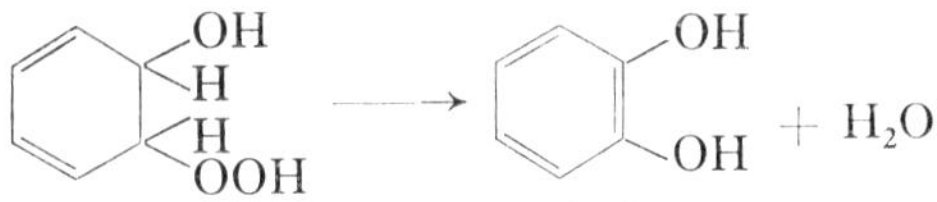

catechol

(6.74)

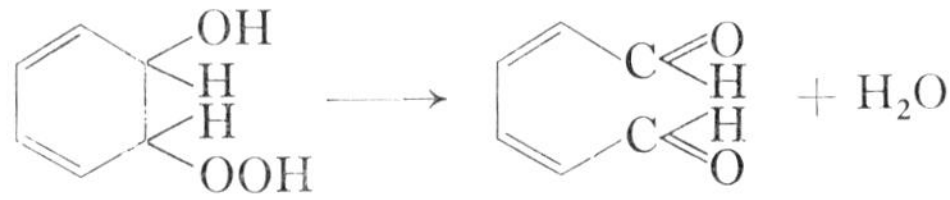

mucondialdehyde (mda)

(6.75)

G(phenol) = 1.9 (prompt) + 0.4 (post-irradiation). G(catechol) = 0.3. G(mucondialdehyde) = 0.8. G(H_2O_2) = 2.2, but this value cannot be derived from the proposed reaction schemes.

In acid, aerated solution the yields are the same, except G(mda) = 1.5 and G(H_2O_2) = 3.2.

There are also variations in yield with dose rate, which usually means that there are radical–radical reactions. The radiation chemistry of benzene in aqueous solution is not completely understood, and some of the above reactions are speculative in that they have not been studied and confirmed by several independent groups.

FURTHER READING

ANBAR, M. and NETA, P., Tables of Bimolecular Rate Constants. *Intern. J. Appl. Radiation Isotopes*, **18**, 493 (1967).

ANBAR, M., Reactions of Hydrated Electrons with Inorganic Compounds, *Quart. Rev. (London)*, **22**, 578 (1968).

† BAXENDALE, J. H., Electrons in Solution, in *Current Topics in Radiation Research*, EBERT, M. and HOWARD, A. (Eds.), North Holland, Amsterdam, Vol. 3 (1967), p. 1.

† DAINTON, F. S., The Simplest Free Radicals, *Pure and Appl. Chem.*, **15** (No. 1), 15 (1966), and The Chemistry of the Electron, in *Proceedings of the Fifth Nobel Symposium—Fast Reactions and Primary Processes in Chemical Kinetics*—CLAESSON, S. (Ed.), Interscience, N.Y. (1967), p. 185.

† DYE, J. L., The Solvated Electron, *Sci. Am.*, **216**, No. 2 (Feb.), 76 (1967).

HART, E. J. (Ed.), Basic Mechanisms in the Radiation Chemistry of Aqueous Media, *Radiation Res.*, Supplement 4 (1964).

HART, E. J. (Ed.), Solvated Electron, Amer. Chem. Soc., *Adv. Chem. Ser.*, No. 50, Washington (1965).

HART, E. J. and ANBAR, M., *The Hydrated Electron*, Wiley, New York (1969).

WALKER, D. C., The Hydrated Electron, *Quart. Rev. (London)*, **21**, 79 (1967).

† These references should be read first.

Chapter 7

Organic Compounds

The susceptibility of organic compounds to radiation varies greatly. Generally, aromatic systems are highly resistant to radiation, although the reasons for this are still the subject of considerable controversy. For example, polyphenyls have been used as coolants in nuclear reactors, because of their radiation resistance. At the other extreme, radiation initiates the polymerization of some olefins and vinyl compounds by chain-growth mechanisms, and G(-monomer) may be as high as 10^6. Chain reactions also occur in radiation-induced oxidations, chlorinations, sulphoxidations and isomerizations. The majority of organic compounds lie between these extremes and undergo radiolysis reactions with total G(product) values of 1 to 10.

In contrast to aqueous solutions, in liquid organic systems there is (1) a low probability of electron escape from the spurs, and (2) a high probability of the primary species reacting with the substrate.

ELECTRON ESCAPE

The probability of escape of the ejected electron from the field of its parent, positive ion is much less than in aqueous solution, i.e. there is greater *geminate recombination* of the ion pairs. Although the primary yield of electrons is $G_e \approx 3$ in all systems, only in aqueous solutions is the readily-measured value as high as 3. Normal measurements have given $G_e \approx 1$ for alcohols and $G_e \approx 0.1$ for hydrocarbons. The trend of these values with the dielectric constant of the medium is shown in Fig. 7.1.

G_{fi} (the radiation yield of free ions) can also be calculated from the electrical conductivity of the liquid under irradiation. The sum of G_{gi} (the yield of geminate ions) and G_{fi} may be evaluated by using scavengers.

It is possible to obtain $G_e \approx 3$ in alcohols if high concentrations of scavengers are used, and in hydrocarbons with still higher concentrations of scavengers. Pulse radiolysis studies on the nanosecond timescale have generally confirmed these observations.

One reason for the greater geminate recombination is that the electron produced by the ionization event, after becoming thermalized, can be solvated only in a polar medium. The lower the dielectric constant, the less is the *trapping* by the dipoles and the freer is the electron to return to its parent, positive ion. Another reason is that the attractive forces between the two oppositely charged ions are greater for materials of lower dielectric constant.

Therefore, reactions involving ions will be relatively unimportant in liquid or solid organic systems, unless ions are formed by such secondary

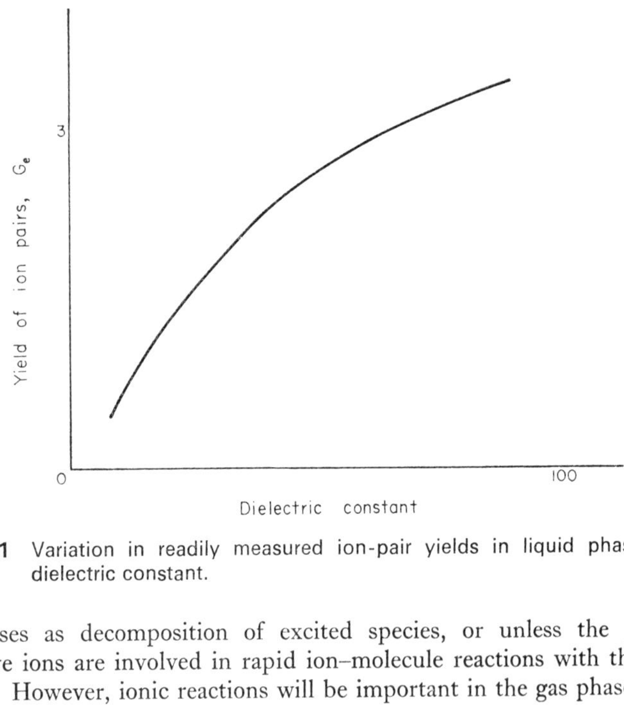

Fig. 7.1 Variation in readily measured ion-pair yields in liquid phase with dielectric constant.

processes as decomposition of excited species, or unless the parent, positive ions are involved in rapid ion–molecule reactions with the substrate. However, ionic reactions will be important in the gas phase.

Excited molecules

When the interaction between ionizing radiation and a molecule is insufficiently energetic to cause ionization, an excited state of the molecule may result. Excited molecules may also be formed in other ways, such as by ion recombination. These excited molecules can undergo some reactions which are impossible in their ground states. They can also transfer energy to other molecules. This is sometimes difficult to demonstrate unequivocally.

Pulse radiolysis studies have shown that in many cases short-lived excited species are found.

Formation of Radicals

Decomposition of neutralized, parent, positive ions and of excited molecules will be the main mode of formation of radicals.

$$M \rightsquigarrow M^+ + e^- \longrightarrow M^* \longrightarrow A\cdot + B\cdot \qquad (7.1)$$

The cage effect will be important in liquids and solids, and only a fraction of the actual bond scissions will lead to radicals with lifetimes greater than a few molecular vibrations.

$$M \rightsquigarrow (A\cdot + B\cdot) \longrightarrow M \qquad (7.2)$$

This effect becomes predominant in the solid state and explains the low yields of radicals trapped in solids irradiated at low temperatures. Small radicals, such as hydrogen atoms, can escape from the cage more readily than large radicals, especially in solids. Therefore, hydrogen atoms will be a major primary product.

RADICAL AND ION–MOLECULE REACTIONS

The radicals will undergo dimerization, abstraction and addition reactions. In contrast to aqueous solutions, where the water is relatively inert, reactions of primary product species with the substrate can be important. These reactions will usually involve abstraction or addition.

The products can be separated by scavenger studies into (*a*) radical, and (*b*) molecular. In this context *molecular* means not produced via scavengable free radicals, and may include—(1) true molecular, (2) ionic, (3) 'hot' radical and (4) radical intra-spur. The distribution of products can usually be explained by either radical or ionic mechanisms. Direct evidence for the presence of free radicals can be obtained from e.s.r. and scavenger studies, whereas ion–molecule reactions are usually inferred from mass spectrometer studies.

Electrons can be scavenged by nitrous oxide, sulphur hexafluoride, carbon dioxide or aromatic compounds; protons by water or ammonia; positive ions and free radicals by nitric oxide; free radicals by iodine, oxygen, D.P.P.H. (1,1-diphenyl-2-picrylhydrazyl), or *p*-benzoquinone; positive ions and electrons by biphenyl, alkyl iodides or T.M.P.D. (tetramethyl-*p*-phenylenediamine).

Unsaturated products are efficient scavengers of free radicals, consequently the *G* values for yields of unsaturates and for products involving radical intermediates are frequently dose-dependent, and subject to considerable doubt. A good example is the controversy about the yield of acetylene from methane; *G* values from zero to 0.7 have been claimed.

The radicals are frequently subject to two competing reactions—a second-order reaction such as dimerization

$$2R\cdot \longrightarrow R_2 \qquad (7.3)$$

and a pseudo-first-order reaction with a major component of the system, e.g. hydrogen abstraction.

$$R\cdot + R'H \longrightarrow RH + R'\cdot \qquad (7.4)$$

Therefore, as the L.E.T. increases, the yield of R_2 will increase and the yield of RH will decrease. The study of L.E.T. effects is providing an insight into the reactions involved in the radiolysis of many liquid organic compounds.

Irradiation of glasses formed by quenching organic compounds at 77°K have revealed much about electron trapping and about the nature of intermediate species.

The main points of contrast between radiolysis in the gas phase and in condensed phases have been given in detail in Chapter 5. Briefly, the yield of primary ion pairs is much greater in the gas phase. With organic systems there are many possibilities for ion–molecule reactions.

TYPICAL EXAMPLES

It is impossible in this short chapter to consider all the organic compounds for which radiolysis products and mechanisms have been investigated. A few systems have been selected to give an indication of what can be expected.

Alcohols

Aliphatic alcohols are moderately polar, hence their radiation chemistry is closely related to that of water. The electron is solvated in alcohols, but not as strongly as in water. Rates of reaction of these solvated electrons with solutes are very nearly the same as those of hydrated electrons. Since aliphatic alcohols dissolve some substances which are sparingly soluble in water, such as hydrocarbons, it is possible to extend our knowledge of the reactions of many compounds with solvated electrons. Once again, rates of reaction are dependent upon the availability of a vacant electron orbital in the molecule. In particular, multi-ring aromatic hydrocarbons, in which the electron can be distributed over two or more rings, react readily with solvated electrons. As a homologous series providing a series of graded properties, the alcohols can reveal other general features of radiation chemical reactions.

Alcohols undergo abstraction reactions with hydrogen atoms.

$$CH_3OH + H\cdot \longrightarrow \cdot CH_2OH + H_2 \qquad (7.5)$$

The radicals formed dimerize to form glycol.

$$2\cdot CH_2OH \longrightarrow \begin{array}{c} CH_2OH \\ | \\ CH_2OH \end{array} \quad (7.6)$$

Formaldehyde is formed principally by a molecular process.

$$CH_3OH^* \longrightarrow HCHO + H_2 \quad (7.7)$$

Alkanes

The main effects of radiation on alkanes are: (1) liberation of H_2, (2) development of unsaturation, (3) formation of dimers, (4) production of fragments from C_1 to C_n and combination of these in pairs. This can be readily seen from the results for *n*-hexane (liquid) shown in Table 7.1.

Table 7.1. Radiolysis products from liquid *n*-hexane (C_6H_{14}).

Product	G Value
H_2	5.0
CH_4	0.1
C_2H_4	0.3
C_2H_6	0.3
C_3H_6	0.1
C_3H_8	0.4
C_4H_{10}	0.5
C_5H_{12}	0.3
C_6H_{12}	1.2
C_7	0.2
C_8	0.5
C_9	0.5
C_{10}	0.4
C_{12}	2.0

Hydrogen and dodecane (dimer) are the main products. There is a complete range of hydrocarbons from C_2 to C_{10}, indicating extensive hydrogen abstraction by, and combination of, fragments. If free radical scavengers, e.g. I_2 or O_2, are present during irradiation, then the hydrogen yield is reduced by 40%, showing that at least this proportion is produced by reactions of hydrogen atoms (radical reactions).

Iodine is frequently used to measure the yields of free radical intermediates. The amounts of alkyl halides produced are measured.

$$R\cdot + I_2 \longrightarrow RI + I\cdot \quad (7.8)$$

However, it is not effective in the solid state. The other halogens cannot be used as radical scavengers, because they also undergo substitution reactions and addition to unsaturated bonds. Nitrous oxide has been used as an electron scavenger.

Radical reactions

The radiolysis can be interpreted as proceeding by scission of C–C and C–H bonds and reaction of the resultant radicals.

$$RH \rightsquigarrow RH^{\bullet +} + e^- \longrightarrow RH^* \longrightarrow R^\bullet + H^\bullet \text{ or } R'^\bullet + R''^\bullet \quad (7.9)$$

$$H^\bullet + H^\bullet \longrightarrow H_2 \quad (7.10)$$

$$R^\bullet + R'H \longrightarrow RH + R'^\bullet \quad (7.11)$$

$$R^\bullet + R'^\bullet \longrightarrow RR' \quad (7.12)$$

Production of unsaturates can be attributed to disproportionation reactions and molecular elimination of hydrogen.

$$C_2H_5^\bullet + C_6H_{13}^\bullet \longrightarrow C_2H_4 + C_6H_{14} \quad (7.13)$$

$$C_6H_{14} \rightsquigarrow C_6H_{12} + H_2 \quad (7.14)$$

The importance of radical reactions following decomposition of the neutralized parent ion is suggested by the estimate, from conductivity measurements, that only 3% of the electrons escape the field of the parent ion in liquid, saturated hydrocarbons.

Ionic reactions

However, it has also been suggested that the time interval before recombination occurs (10^{-7} seconds) is greater than the time required for ion–molecule reaction to occur. Thus, the radiolysis could be explained by reaction of the parent ion with substrate, or reaction of its decomposition products.

$$C_6H_{14}^{\bullet +} + C_6H_{14} \longrightarrow C_6H_{15}^+ + C_6H_{13}^\bullet \quad (7.15)$$

$$C_6H_{14}^{\bullet +} \longrightarrow C_2H_5^+ + C_2H_4 + C_2H_5^\bullet \quad (7.16)$$

$$C_6H_{14}^{\bullet +} \longrightarrow C_4H_8^+ + C_2H_6 \quad (7.17)$$

Direct elimination of hydrogen, and sometimes other small molecules, from positive ions has also been proposed.

$$C_6H_{14}^{\bullet +} \longrightarrow C_6H_{12}^{\bullet +} + H_2 \quad (7.18)$$

The ion–molecule reactions are considered to involve transfer of (1) hydride ion (H^-), (2) hydrogen atom ($H^\bullet$), (3) hydrogen molecule (H_2),

(4) hydrogen ion ($H_2^{\cdot -}$), (5) proton (H^+) or (6) charge, i.e. an electron (e^-). For example:

$$(H^-)\quad C_2H_5^+ + C_6H_{14} \longrightarrow C_2H_6 + C_6H_{13}^+ \qquad (7.19)$$

$$(H_2)\quad C_6H_{14}^{\cdot +} + C_2H_4 \longrightarrow C_2H_6 + C_6H_{12}^{\cdot +} \qquad (7.20)$$

$$(H_2^{\cdot -})\quad C_4H_8^{\cdot +} + C_6H_{14} \longrightarrow C_4H_{10} + C_6H_{12}^{\cdot +} \qquad (7.21)$$

$$(e^-)\quad C_4H_8^{\cdot +} + C_6H_{14} \longrightarrow C_4H_8 + C_6H_{14}^{\cdot +} \qquad (7.22)$$

Some evidence for ion–molecule reactions can be obtained by using electric field techniques (although additional effects due to the field may occur). Recently, ammonia has been suggested as a scavenger for positive ions. The increase in yield of unsaturated products should be equal to the yield of carbonium ions. Thus—

$$C_3H_7^+ + NH_3 \longrightarrow C_3H_6 + NH_4^+ \qquad (7.23)$$

There is some indication that in hydrocarbon chains C–C scission is favoured 2–4 bonds from the end of the chain. Branched structures have a higher probability of rupture than linear chains and tertiary carbon atoms are particularly sensitive. For example, the methane yield increases rapidly in the series *n*-hexane, 2-methyl pentane, 2,2-dimethyl butane.

Alkyl halides

Alkyl halides are quite sensitive to radiation, but there is considerable variation in the products formed. This can be correlated to some extent with the particular halide. There is a gradation in bond strength from C–I through C–Br and C–Cl to C–F. The C–I bond is much weaker than C–C or C–H, whereas C–F is much stronger, as shown in Table 7.2. Consequently, iodine-atom abstraction takes place, but not fluorine-atom abstraction—hydrogen-atom abstraction occurring instead.

Table 7.2. Bond strengths in alkyl halides.

Bond	C–I	C–Br	C–Cl	C–F	C–C	C–H
Bond strength						
kcal/mol	46	54	78	107	80	103
kJ/mol	192	226	326	448	334	431

Another feature of the radiolysis of alkyl halides is that isomerization may occur, e.g. *n*-butyl chloride to *sec*-butyl chloride and *n*-propyl chloride to *iso*propyl chloride. The very high yields of isomer indicate a chain reaction, and the rearrangement has been suggested to occur in the alkyl radical to a more stable radical, via the following reaction sequence.

$$n\text{-PrCl} \rightsquigarrow n\text{-Pr}\cdot + \text{Cl}\cdot \qquad (7.24)$$

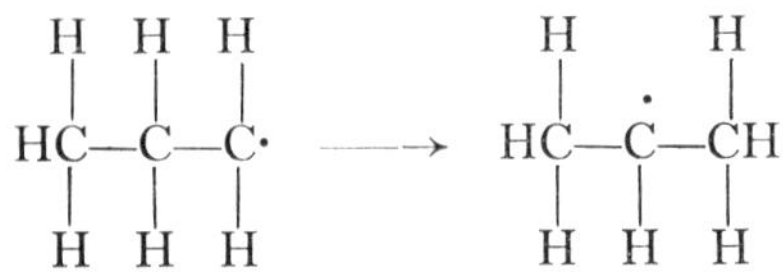

n-Pr· → *iso*-Pr· (7.25)

$$iso\text{-Pr}\cdot + n\text{-PrCl} \longrightarrow iso\text{-PrCl} + n\text{-Pr}\cdot \qquad (7.26)$$

Most radiolysis studies have been made on alkyl iodides, and the main products from liquid ethyl iodide are shown in Table 7.3.

Table 7.3. Radiolysis of liquid ethyl iodide C_2H_5I.

Product	G value
H_2	0.2
I_2	2.1
HI	0.3
CH_4	0.01
C_2H_4	2.2
C_2H_6	1.9
C_4H_{10}	0.3

Halogen is a major product from alkyl iodides, but the hydrogen halide becomes important in the chlorides. Neither are to be expected in the fluorides to any extent. Results from the radiolysis of *n*-butyl chloride are given in Table 7.4.

Table 7.4. Radiolysis of *n*-butyl chloride.

Product	G value
H_2	1.4
Cl_2	0
HCl	1.0
n-C_4H_{10}	6.9
sec-butyl chloride	60
dichlorobutanes	1.5

Radiolysis mechanism

The mechanism of radiolysis of alkyl halides is still largely unresolved. As usual, free radical and ionic reactions have been proposed. Support for a free radical mechanism has come from similarities between the radiolysis and photolysis of alkyl iodides. However, this may simply reflect the weakness of the C–I bond in any mechanism. Free radicals are undoubtedly involved. Alkyl radicals have been detected in irradiated alkyl iodide, bromide and chloride glasses and polycrystalline solids by e.s.r. Therefore, a primary reaction could be homolytic scission of the C–I, C–Br or C–Cl bond, e.g.

$$CH_3I \rightsquigarrow CH_3\cdot + I\cdot \qquad (7.27)$$

There is also evidence that appreciable recombination of geminate alkyl radicals and iodine atoms occurs—an example of the cage effect.

$$RI \rightsquigarrow (R\cdot + I\cdot) \longrightarrow RI \qquad (7.28)$$

The radiolysis has also been interpreted in terms of carbonium ion–molecule reactions by comparison with mass spectrometer studies.

However, alkyl halides have high electron affinities and for weak C–X bonds dissociative electron capture would be highly favoured.

$$RI + e^- \longrightarrow R\cdot + I^- \qquad (7.29)$$

Visible-ultraviolet and e.s.r. spectra attributed to negatively and positively charged species have been observed in irradiated glasses at 77°K, but their specific identification is still in question.

Photolysis studies on gaseous and liquid ethyl iodide suggest that HI may be formed by a true molecular reaction.

$$C_2H_5I \longrightarrow C_2H_4 + HI \qquad (7.30)$$

There is evidence that in the radiolysis of liquid alkyl iodides the relative amounts of HI and H_2 are controlled by competition between the reactions.

$$R\cdot + HI \longrightarrow RH + \tfrac{1}{2}I_2 \qquad (7.31)$$

$$R\cdot + I_2 \longrightarrow RI + \tfrac{1}{2}I_2 \qquad (7.32)$$

Quite different yields of HI and I_2 have been obtained from alkyl iodides in the solid and liquid phases, as shown in Table 7.5

Table 7.5. Effect of phase on radiolysis of alkyl iodides.

Compound	G(HI) Liq	G(HI) Glass	G(HI) Cryst	$G(I_2)$ Liq	$G(I_2)$ Glass	$G(I_2)$ Cryst
n-C_3H_7I	0.1	2.6	2.3	1.5	0.5	0.5

Similar results have been obtained for other iodides. The reason for these differences has not yet been established, but it probably results from reduced diffusion of radicals in the solid state.

Olefins

Irradiation of olefins leads principally to the formation of low-molecular-weight polymers. Olefins do not normally polymerize by a free radical mechanism (ethylene is an exception) and ionic reactions are postulated. The polymers produced are mainly mono-olefinic and the bonds linking the monomer units are formed between carbon atoms which were originally joined by double bonds in the monomer. The mechanism proposed is—

$$C_nH_{2n} \rightsquigarrow C_nH_{2n}{}^{\cdot +} + e^- \qquad (7.33)$$

$$C_nH_{2n}{}^{\cdot +} + C_nH_{2n} \longrightarrow C_{2n}H_{4n}{}^{\cdot +} \qquad (7.34)$$

$$C_{2n}H_{4n}{}^{\cdot +} + e^- \longrightarrow C_{2n}H_{4n} \qquad (7.35)$$

Other products include—

(1) cis–trans isomer

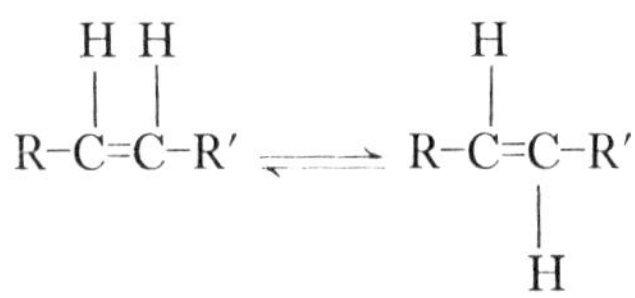

(7.36)

(2) alkane, resulting from saturation of the double bond.

$$C_nH_{2n} \longrightarrow C_nH_{2n+2} \qquad (7.37)$$

(3) alkynes or diolefins, formed by loss of hydrogen atoms from adjacent carbon atoms—probably by molecular elimination.

$$R{-}CH{=}CH{-}R' \longrightarrow R{-}C{\equiv}C{-}R + H_2 \qquad (7.38)$$

$$R{-}CH_2{-}CH_2{-}CH{=}CH{-}R' \longrightarrow R{-}CH{=}CH{-}CH{=}CH{-}R' + H_2 \qquad (7.39)$$

(4) adjacent olefin, due to hydrogen-atom migration

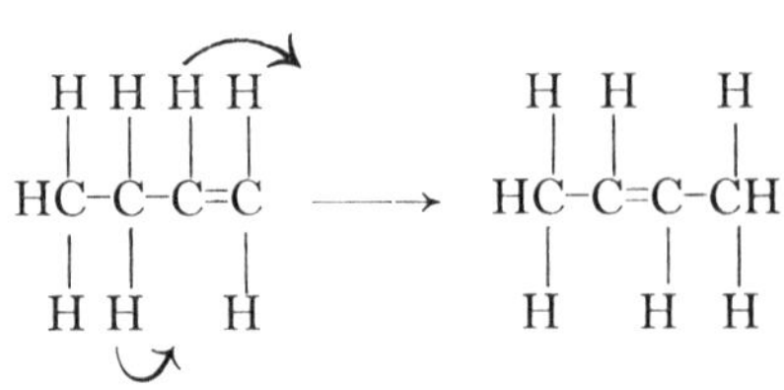

(7.40)

(5) skeletal isomer—this is much more important in branched than straight-chain olefins and includes cyclopropane derivatives.

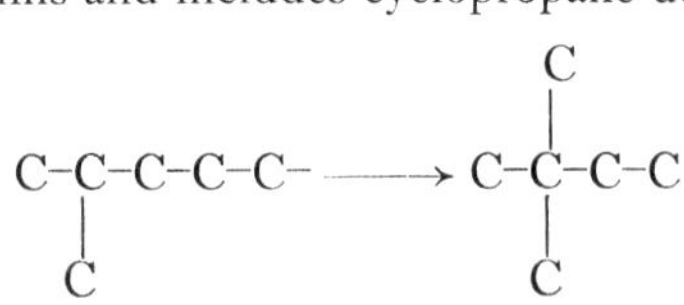

(7.41)

Also, the polymers from branched olefins are mainly skeletally isomerized.

(6) fragmentary products—these are usually obtained in small yield, on account of the scavenging ability of the double bonds for radicals.

Vinyl compounds may be considered as a class of substituted olefins. In many cases long-chain polymers are formed on irradiation, and this may occur by free radical or ionic mechanisms. Radiation-induced polymerization is considered in more detail in Chapter 8.

Aromatics

Aromatic compounds show a relatively high resistance to radiation. In particular, the yields of gaseous products are small. For benzene the values given in Table 7.6 have been obtained.

Table 7.6. Radiolysis of benzene.

$G(-C_6H_6)$	0.8
$G(H_2)$	0.04
$G(C_2H_2)$	0.02

Two factors have been advanced to explain the low product yields from aromatic compounds:

(1) the ability of the conjugated ring system to degrade excitation energy without concentration of energy in one bond and consequent scission.

(2) efficient scavenging of radical fragments by the aromatic ring, instead of hydrogen abstraction occurring.

Addition, instead of abstraction, has been demonstrated in the gas phase by pulse radiolysis of mixtures of benzene and hydrogen. The pulses produce hydrogen atoms very rapidly, and the major subsequent products from the benzene are 1,4-cyclohexadiene and dicyclohexadiene, attributed to the reaction sequence—

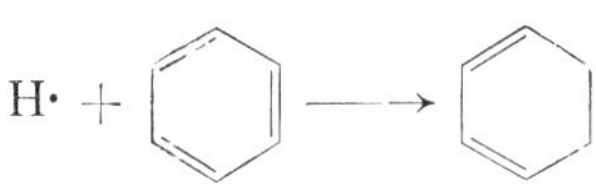
cyclohexadienyl radical (7.42)

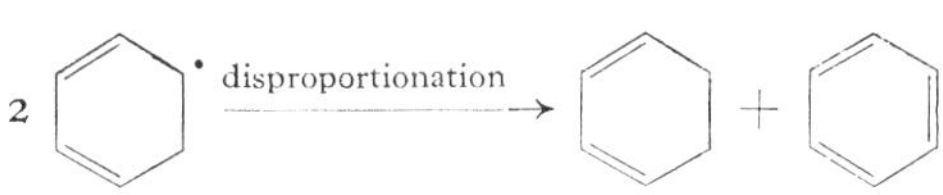

1,4-cyclohexadiene
+ benzene (7.43)

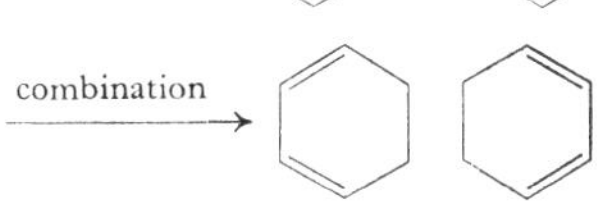

di-cyclohexadiene (7.44)

Hydrogen abstraction to produce the phenyl radical ($C_6H_5\cdot$) does not occur to any extent, since the yield of diphenyl is negligible.

The cyclohexadienyl radical ($C_6H_7\cdot$) has been identified in gas-phase pulse radiolysis, in radiolysis of low-temperature glasses and on addition of an H• atom to benzene in liquid systems.

Alkyl substitution increases the radiation sensitivity, but there is still much less abstraction than addition and gas yields are low, relative to similar saturated compounds.

Scission of bonds to the aromatic carbon atoms is not favoured. For example, when ethyl toluene is irradiated, methane is the main product ($G = 0.02$) after hydrogen ($G = 0.2$), and 85% of it comes from rupture of the C–C bond in the ethyl group.

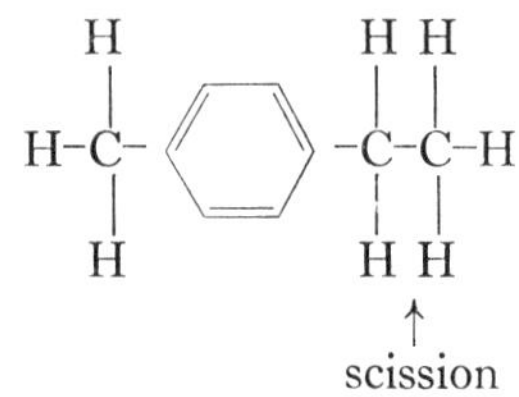

The hydrogen appears to be produced by a 'molecular' reaction.

One aromatic compound in which high radiation sensitivity has been reported is thiophenol, C_6H_5SH. The major products are H_2 ($G = 4.2$) and C_6H_5S–SC_6H_5 ($G = 4.6$). This seems to be an exceptional case, due to the special liability of the C–S bond to rupture.

RADIATION PROTECTION

When a homogeneous mixture of two compounds is irradiated, the yields of the different products may be expected to be directly proportional to the yields from the pure components and their relative proportions in the mixture. These proportions must be the electron fractions. This expectation is based on the assumption that each component is degraded independently of the other. However, in some cases there are marked deviations from this expectation. The classical example is in mixtures of cyclohexane and benzene, as shown in Fig. 7.2.

Most work has been done on the gaseous products, but other products show this type of behaviour.

The explanation of this *protective effect* of the benzene on the cyclohexane remains highly controversial. Theories of energy transfer from cyclohexane to benzene have been advanced. This could be by excitation or charge transfer. The latter phenomenon has been unambiguously shown to occur in the gas-phase radiolysis of mixtures of argon or krypton with simple hydrocarbons. The large number of excited states in aromatic systems supports the feasibility of excitation transfer. Either process is a non-chemical effect, involving a precursor of cyclohexane decomposition.

On the other hand, chemical scavenging of radical intermediates by the benzene or by olefins has been proposed as an explanation. In fact, a great difficulty in determining the product yields from cyclohexane, pure or with small amount of benzene, is the dose-dependence at low doses. This indicates radical scavenging by olefins, believed to be mainly cyclohexene, produced from the cyclohexane. The most likely radicals to be scavenged are hydrogen atoms.

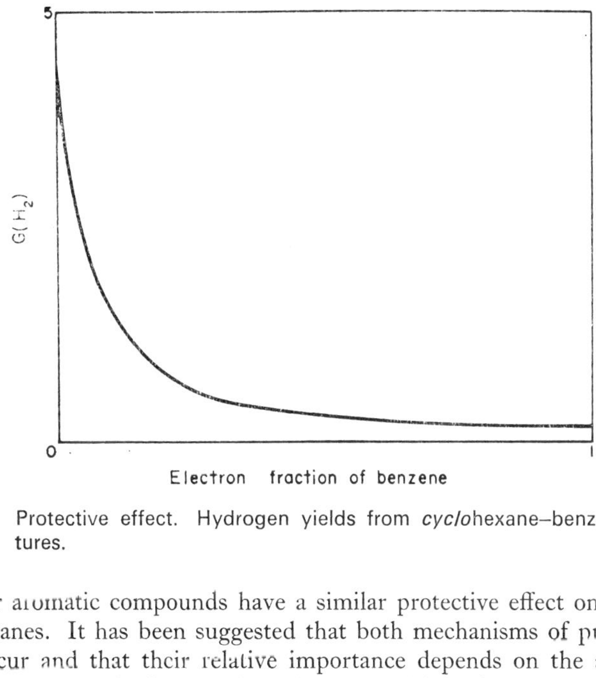

Fig. 7.2 Protective effect. Hydrogen yields from *cyclo*hexane–benzene mixtures.

Other aromatic compounds have a similar protective effect on various cycloalkanes. It has been suggested that both mechanisms of protection may occur and that their relative importance depends on the aromatic compound, the cycloalkane and on the composition of the mixture.

It is also interesting to note that the product yields from mixtures of aryl cycloalkanes with benzene showed a linear relationship with composition. Apparently these compounds are internally protected and the benzene acts only as a diluent.

Another example of a type of radiation protection occurs in mixtures containing thiol compounds. These are particularly liable to attack by radicals, with scission of the S–H bond and formation of a sulphuryl radical.

$$R\cdot + R'SH \longrightarrow RH + R'S\cdot \qquad (7.45)$$

The sulphuryl radicals are relatively stable, do not abstract hydrogen atoms and usually disappear by combination to form a disulphide linkage.

$$2RS\cdot \longrightarrow RSSR \qquad (7.46)$$

Their protective effect has been mainly investigated in biological systems, because many proteins and enzymes contain sulphide and disulphide linkages.

FURTHER READING

BOLT, R. O. and CARROL, J. G. (Eds.), *Radiation Effects on Organic Materials*, Academic, New York (1963).

BURNS, W. G. and BARKER, R., Dose-rate and Linear Energy Transfer Effects in Radiation Chemistry in PORTER, G. (Ed.), *Progress in Reaction Kinetics*, Vol. 3, Pergamon, Oxford (1965).

GAUMANN, T. and HOIGNÉ, J. (Eds.), *Aspects of Hydrocarbon Radiolysis*, Academic, New York (1968).

KEVAN, L. and LIBBY, W. F., Ionic States in Solid Hydrocarbons, in NOYES, W. A., HAMMOND, G. S. and PITTS, J. N. (Eds.), *Advances in Photochemistry*, Interscience, New York (1964).

PHILLIPS, G. O. (Ed.), *Energy Transfer in Radiation Processes*, Elsevier, Amsterdam (1966).

† SHERMAN, W. V., Free-Radical Intermediates in the Radiation Chemistry of Organic Compounds, in WILLIAMS, G. H. (Ed.), *Advances in Free Radical Chemistry*, Vol. 3, Academic, New York (1969).

SWALLOW, A. J., *Radiation Chemistry of Organic Compounds*, Pergamon, Oxford (1960).

TOPCHIEV, A. V. (HOLROYD, R. A., Ed.), *Radiolysis of Hydrocarbons*, Elsevier, Amsterdam (1964).

WILLIAMS, F., Ionic Processes in Radiolysis of Hydrocarbons, *Quart. Rev. (London)*, **17**, 101 (1963).

† This reference should be read first.

Chapter 8

Monomers and Polymers

Small amounts of radiation can produce relatively large effects in monomers or polymers. The radiation initially produces ions and radicals with radiation chemical yield (G value) of 1–10, as in other systems. These ions and radicals may initiate polymerization or copolymerization in monomers and degradation or crosslinking in polymers.

A single radical or ion can initiate a polymerization chain reaction resulting in addition of 10^3–10^5 monomer units. Thus there is a very large consumption of monomer, i.e. a high G(−monomer).

Ionization or radical formation in a substrate polymer can initiate *graft copolymerization* of another monomer having different properties to give a product combining desirable features of both.

Radiation may produce dramatic changes in the properties of polymers as a result of (1) *cleavage* (or *scission*) of main chain bonds, or (2) *crosslinking* of the polymer molecules. Because of the large size of polymer molecules, a small number of such reactions will affect a relatively large proportion of the molecules in a sample of polymer.

The comparatively large radiation-induced changes in monomer–polymer systems make them attractive subjects for industrial irradiation processes and a number are already in operation.

Biological macromolecules are very sensitive to radiation. Increasing knowledge of their structures can be expected to lead to greater understanding of the specific effects of radiation. Protection against radiation damage may then become possible.

MONOMERS

Monomers which polymerize by addition chain reactions can be polymerized by radiation. The reaction sequence is—

initiation:

$$M \rightsquigarrow M^{\ddagger} \qquad (8.1)$$

propagation:

$$M^{\ddagger} + M \longrightarrow M_2^{\ddagger} \qquad (8.2)$$

$$M_i^{\ddagger} + M \longrightarrow M_{i+1}^{\ddagger} \qquad (8.3)$$

termination:

$$M_i^{\ddagger} + M_j^{\ddagger} \longrightarrow P_{i+j} \text{ or } P_i + P_j \qquad (8.4)$$

$$M_i^{\ddagger} + X \longrightarrow P_i + X^{\ddagger} \qquad (8.5)$$

where M represents a monomer molecule, $M^{\ddagger}$ an active centre produced by irradiation, P a polymer molecule and X a *chain terminating* additive. $M^{\ddagger}$ may be a free radical or an ion.

Many cyclic compounds or those containing double bonds (especially vinyl compounds) polymerize in this way. Free radical, cationic and anionic initiating active centres have been demonstrated in radiolysis. Radicals are produced by homolytic bond cleavage,

$$M \rightsquigarrow A\cdot + B\cdot \qquad (8.6)$$

where one of the radical fragments is frequently a hydrogen atom, or by addition of a hydrogen atom to the double bond of the monomer, e.g.

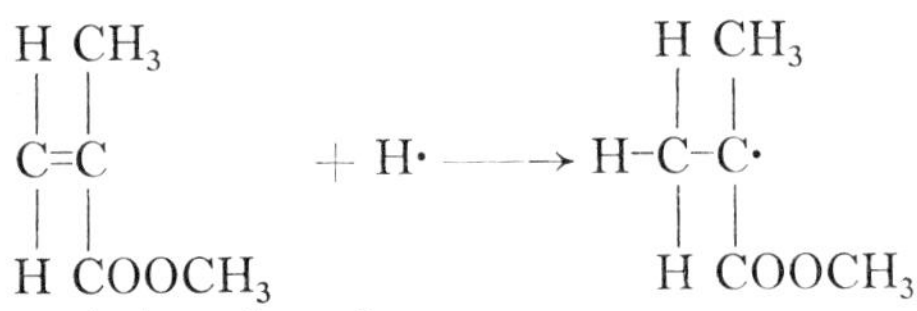

methyl methacrylate (8.7)

Cationic initiating species result from heterolytic bond scission,

$$M \rightsquigarrow A^+ + B^- \qquad (8.8)$$

or electron ejection, followed by hydrogen atom transfer,

$$M \rightsquigarrow M\cdot^+ + e^- \qquad (8.9)$$

$$M\cdot^+ + M \longrightarrow M^+ + M\cdot \text{ or } {}^+M\text{—}M\cdot \qquad (8.10)$$

or by addition of a positive ion, particularly a proton, to the double bond, e.g.

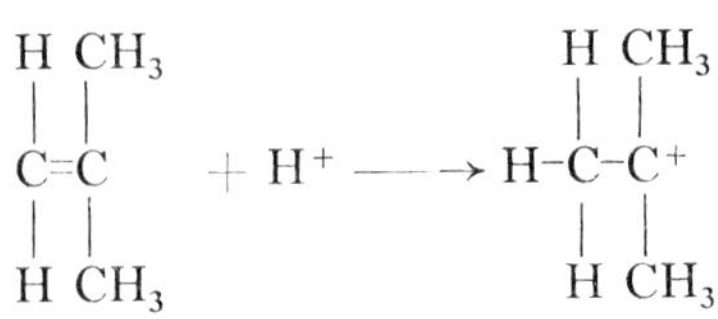

isobutene (8.11)

There is evidence that the cation–radical can also initiate polymerization.

Some monomers, e.g. vinyl halides ($CH_2 = CHX$) polymerize only by a free radical mechanism, whereas others, like isobutene, polymerize only by a cationic mechanism. Therefore, it is not difficult to demonstrate unambiguously the presence of free radicals or cations during radiolysis. If the amount of polymer formed and its molecular weight (number average, $\overline{M}_n$) are determined, then the number of polymer molecules can be calculated. This gives the number of initiating centres, e.g. free radicals, provided the mechanism of termination is known, unless chain transfer occurs. *Chain transfer* is reaction of a growing chain with a compound Y to terminate the first chain, but with $Y^{\ddagger}$ able to start another polymerization chain reaction.

$$M_i^{\ddagger} + Y \longrightarrow P_i + Y^{\ddagger} \qquad (8.12)$$

$$Y^{\ddagger} + M \longrightarrow YM^{\ddagger} \qquad (8.13)$$

It occurs readily with many solvents and also with monomer and with polymer.

Polymerization of some monomers can proceed by an anionic mechanism and anions may be formed during radiolysis by electron capture and electron transfer,

$$M + e^- \longrightarrow M^{\cdot -} \qquad (8.14)$$

$$M^{\cdot -} + M \longrightarrow A^- + B^{\cdot} \text{ or } {}^-M\text{–}M^{\cdot} \qquad (8.15)$$

or by heterolytic bond scission (equation *8.8*). Most monomers which can be polymerized by an anionic mechanism can also be polymerized by a cationic or radical mechanism, so that evidence for radiation-induced anionic polymerization is difficult to obtain.

Radiation-induced anionic polymerization of nitroethylene has been demonstrated. Using anion scavengers, a relatively high *G*(initiation) value of about three was obtained, compared with values in the range 0.1–1 for cationic initiation.

Free radical polymerization can be confirmed by observation of e.s.r. spectra of the propagating radicals. Steady state concentrations are normally below detection levels, but higher concentrations can be obtained by trapping in gels or frozen solids. Yields of initiating free radicals in irradiated monomers have been found to be in the range $G = 0.5 - 10$. Additives may indicate the mode of polymerization. Oxygen is a very efficient radical scavenger, causing induction periods, until the oxygen is used up, before polymerization proceeds. Similarly ionic reactions can be affected, e.g. by water or dimethyl formamide. The role of electrons in the radiolysis of monomers can be shown by the profound effects of additives, such as solid zinc chloride, which alter the electron concentration. When a mixture of two monomers is irradiated, the composition of the polymer formed from different co-monomer compositions can be a sensitive test of the reaction mechanism, as shown in Fig. 8.1. Positive and negative styrene ions have been observed in suitable glasses by visible-ultraviolet and e.s.r. spectroscopy and in both the liquid and solid phases by pulse radiolysis.

Studies with rigorously dried monomers indicate that in the presence of water at concentrations higher than 10^{-10} M, radiation-induced cationic polymerization proceeds through an ion-pair, whereas at lower concentrations it occurs via the free ion. The rate of polymerization via the

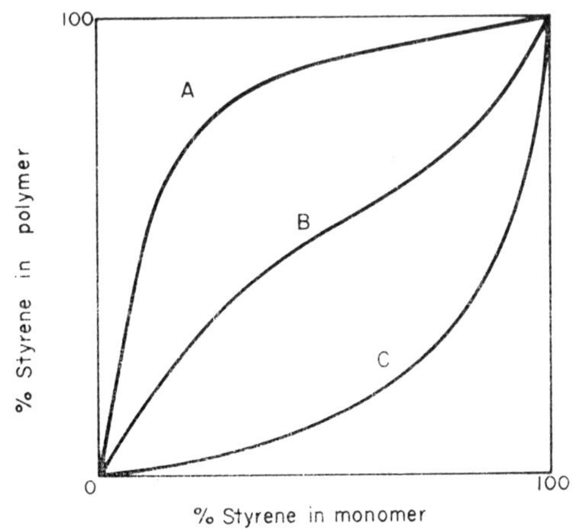

Fig. 8.1 Variation in polymer composition with co-monomer composition for methyl methacrylate–styrene mixtures polymerized by different mechanisms. (A) cationic, (B) free radical, (C) anionic.

free ion is very much greater than for the ion-pair (a factor of 10^6 for styrene) and the steric structure of the polymer is also different.

Polymer formation during irradiation of liquid monomer follows a *sigmoidal* (**S**-shaped) curve with time. This is illustrated in Fig. 8.2. The rate of polymerization increases with dose rate, but may not show much temperature dependence. This is because the *initiation rate* (production of radicals or ions) is usually independent of temperature and the increase in *propagation rate* at higher temperatures is opposed by the increase in *termination rate*. Ionic polymerizations are frequently best carried out at low temperatures.

Radiation may be used to initiate all types of polymerization processes. These include gas phase, heterogeneous and emulsion polymerizations. There is considerable interest in radiation-induced polymerization of aqueous emulsions because the heat of reaction is removed by the water, the molecular weight of the polymer is high and the product is free of residual chemical initiator. However, there appears to be a wide variety of kinetic schemes applicable to different monomers.

Also, a number of monomers can be copolymerized into the same polymer molecules. A copolymerization of particular interest is the gas phase, free radical addition of ethylene and carbon monoxide to form a poly(ketone).

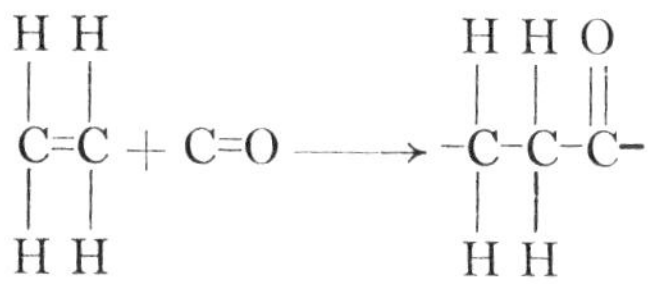

(8.16)

The CO content of the copolymer depends on the composition of the co-

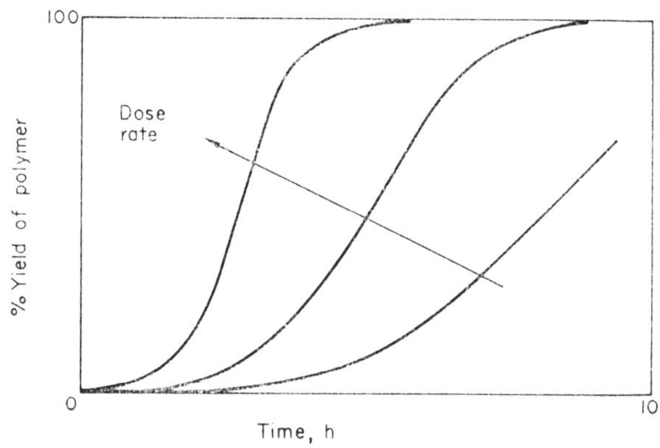

Fig. 8.2 Polymerization of a monomer during irradiation.

monomer mixture and the temperature of polymerization, but is typically 30–50 mole %.

In addition to polymerization, a variety of other chemical reactions occur during irradiation of monomers. These reactions are similar to those observed in comparable organic compounds which do not polymerize and are discussed in Chapter 7.

Solid-State Polymerization

Recently interest has developed in radiation-induced polymerization of *solid* monomers. This is partly due to increasing concern with solid-state processes, and partly in the hope of producing new or improved polymers. Free radical polymerization has been shown to occur in acrylamide and many methacrylates, whereas trioxane undergoes cationic polymerization.

The conversion curves for polymerization during irradiation are similar to those shown in Fig. 8.2.

Post-irradiation polymerization may also occur. The monomer can be irradiated at a low temperature and then heated to a suitable temperature for polymerization. Typical kinetic behaviour is shown in Fig. 8.3. The rate of polymerization is greatest initially and declines to a very low value at a 'limiting' conversion, usually well below 100%. Both the rate and the limiting conversion normally increase with radiation dose and with polymerization temperature.

Copolymerization has been shown to occur in the solid state. The highest rates are often observed in *eutectic* mixtures, when very small crystals are intimately mixed, or in *solid solutions.* Random copolymers are usually formed from solid solutions and block copolymers from eutectic mixtures.

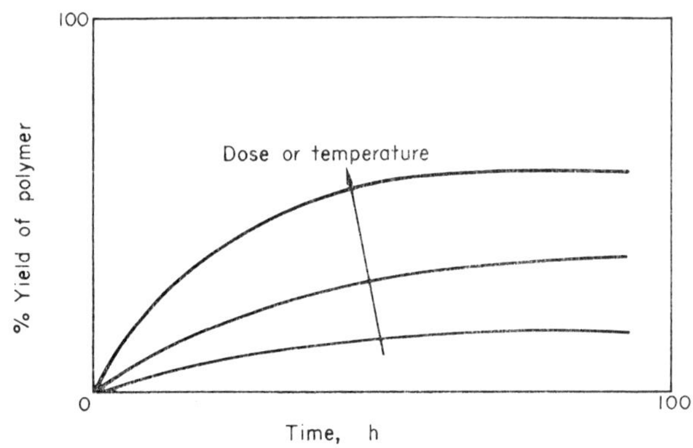

Fig. 8.3 Post-irradiation polymerization in the solid-state.

Additives may, or may not, have an effect on solid-state polymerization.

The *crystalline structure* of the monomer has an important role in solid-state polymerization. It has been shown that the rate of polymerization is dependent on suitable proximity and arrangement of the monomer molecules. Such *topochemical* effects have been observed previously in other systems. Furthermore, the crystalline structure of the polymer product may be directly related to that of the monomer. This *topotaxy* occurs in the polymerization of many ring compounds, such as trioxane.

Another important factor is *molecular mobility.* For example, methyl methacrylate polymerizes only with difficulty (low yields after high doses) in the pure solid-state, but solid solutions in paraffin oil polymerize readily. In many monomers there is a rapid increase in polymerization rate as the melting point is approached. This is probably partly due to the increasing disorder and associated mobility.

A serious limitation in solid-state polymerization is the difficulty of removing the heat of reaction, and explosions can occur.

POLYMERS

The effects of radiation on polymers include the formation of gaseous products, the reduction of existing unsaturation and the production of new unsaturation. But the two reactions which cause the major alterations in the properties of a polymer are scission of main-chain bonds (degradation) and chemical bonding between different polymer molecules (cross-linking).

Polymers have been classified into two groups according to their behaviour under irradiation—those which predominantly *crosslink* and those which *degrade*. For vinyl-type polymers degradation usually occurs when there is a tetra-substituted carbon atom in the main-chain, e.g.

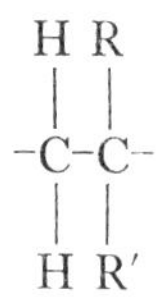

A—degrade

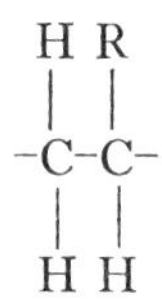

B—crosslink

Thus poly(methyl methacrylate) and poly(isobutene) degrade, but poly(ethylene) crosslinks. However, there are many exceptions to this rule. It has been suggested that a more fundamental factor is the free energy of propagation. When this is low, steric and electronic factors favour degradation. Usually, both degradation and crosslinking occur and the ratio is affected by temperature, crystallinity, stereoregularity and air.

Degradation

Oxygen promotes degradation, apparently by the formation of peroxy species, which prevent the permanent recombination of the geminate radical ends.

Polymers which degrade include poly(isobutene), poly(methyl methacrylate), poly(tetrafluoroethylene), poly(α-methyl styrene) and cellulose. Poly(styrene) degrades when irradiated in air, but crosslinks in vacuum.

The molecular weight of the polymer decreases exponentially with dose (Fig. 8.4). It can be measured readily from the solution properties, provided the polymer has a suitable solvent. The methods include osmometry (number average, $\overline{M}_n$) and light scattering (weight average, $\overline{M}_w$), but most conveniently viscosity (viscosity average, $\overline{M}_v$). If $1/\overline{M}_n$ is plotted against dose a linear relationship is obtained. The slope is the radiation chemical yield for main-chain breaks or scission, G (scission). Corrections must be made if the initial molecular-weight distribution of the polymer is not the most probable, or normal, distribution. This distribution is produced by many polymerization mechanisms, e.g. by free radical addition (*a*) in the presence of adequate chain transfer agent, or (*b*) with

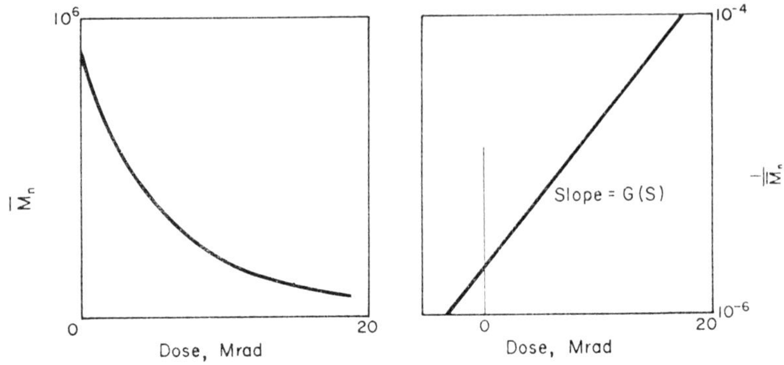

Fig. 8.4 Variation of molecular weight with radiation dose for a degrading polymer.

termination by disproportionation. A normal molecular weight distribution (and the relationship of $\bar{M}_n$, $\bar{M}_w$ and $\bar{M}_v$) is shown in Fig. 8.5. Random scission produces such a normal distribution from any initial distribution. Deviation of the initial polymer from a normal distribution is shown by a negative intercept, which is sometimes called a virtual dose.

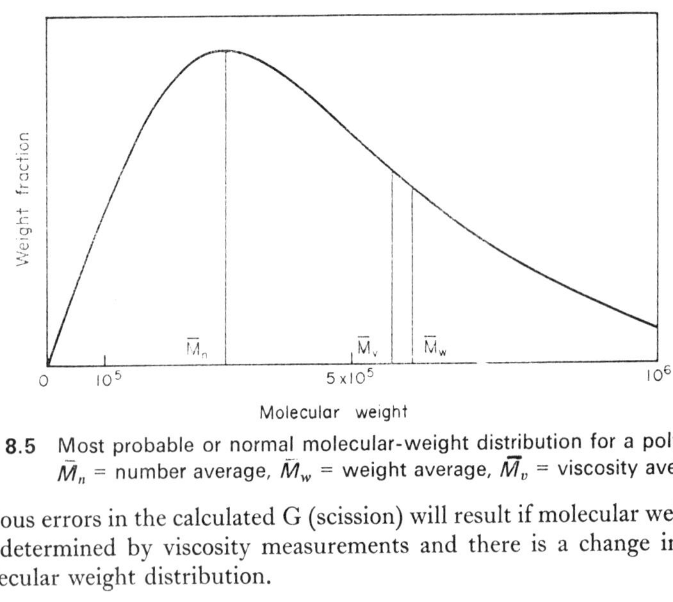

Fig. 8.5 Most probable or normal molecular-weight distribution for a polymer. $\bar{M}_n$ = number average, $\bar{M}_w$ = weight average, $\bar{M}_v$ = viscosity average.

Serious errors in the calculated G (scission) will result if molecular weights are determined by viscosity measurements and there is a change in the molecular weight distribution.

The G (scission) values for degrading polymers are shown in Table 8.1. The value for cellulose is much higher than for the others, probably due to specific cleavage of weak, inter-unit linkages. The temperature of irradiation does not generally have much effect on the amount of scission.

Table 8.1. G (scission) at 20°C for polymers which degrade under irradiation.

Poly(isobutene)	4
Poly(methyl methacrylate)	2
Poly(α-methyl cellulose)	16
Poly(tetrafluoroethylene)	—
Cellulose	11
Poly(α-methyl styrene)	0.3
Poly(ethylene oxide)	—

If main-chain bonds in polymer molecules are broken during irradiation, either directly or as a result of subsequent reactions, the possibility of recombination would be expected to be high, especially in the solid-state. The Franck–Rabinowitch cage effect would be very effective when such large fragments are involved. Early workers considered that only C–H bonds could be permanently broken during irradiation of polymers. Indeed, it has been shown that the number of other permanent breaks is much smaller than the number of repaired breaks.

When degradation of the polymer is predominant, structural strength and plasticity are rapidly lost. The polymer may actually crumble to a powder. The effect of radiation on some polymers is shown in Plate 6. Such diminished physical properties are particularly noticeable in poly(tetrafluoroethylene), which, when unirradiated, shows great strength and resistance to chemical effects over a wide temperature range.

Crosslinking

Crosslinking of polymer chains is equivalent to dimerization in smaller molecules. Just one crosslink attached to each molecule will produce a three-dimensional structure. This results in insolubility, increased strength and elimination of the melting point. These desirable effects have led to industrial interest. Plate 7 shows how a poly(ethylene) bottle given 30 Mrad of γ-irradiation is unaffected by heating in an oven at 120°C, whereas an identical bottle, which has not been irradiated, collapses.

The mechanism of crosslinking is the subject of considerable controversy and probably varies between different polymers. Three main radical processes have been suggested.

(1) Cleavage of a C–H bond on one polymer chain to form a hydrogen atom, followed by abstraction of a second hydrogen atom from a neighbouring chain to produce hydrogen. Then the two adjacent polymeric radicals combine to form the crosslink,

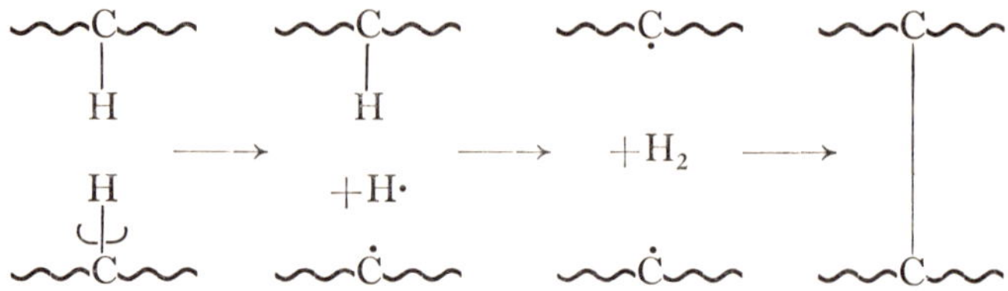 (8.17)

(2) Migration of radical sites, produced by C–H bond cleavage, along the polymer chains until two of them are adjacent, when they combine to form the crosslink.
(3) Reaction of unsaturated groups with hydrogen atoms to form polymeric radicals which can then combine.

Ionic mechanisms may be important, especially at low temperatures, when hydrogen atom abstractions must be precluded by activation energy

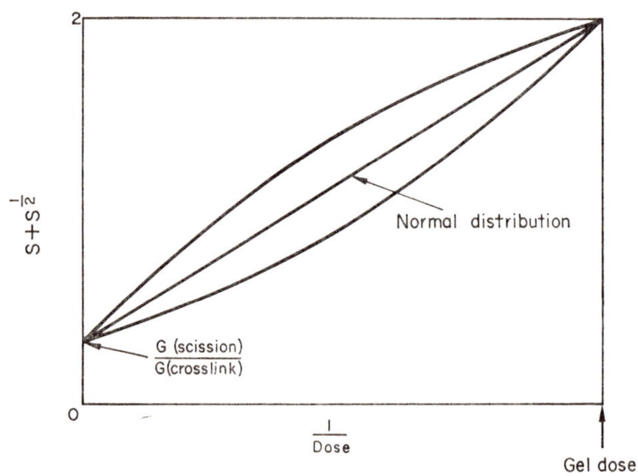

Fig. 8.6 Determination of gel dose and ratio of G(scission) to G(crosslink) for irradiated polymer. S = soluble fraction.

considerations, unless they occur with excited species. However, ionic reactions are difficult to study in such complex, solid materials.

Crosslinking is readily demonstrated by the onset of partial insolubility, i.e. gel formation. The radiation dose required is known as the gel dose and is a measure of the crosslinking yield.

Crosslinking is frequently accompanied by some degradation (scission), and the ratio of G (scission) to G (crosslink) can be obtained from a plot of the type shown in Fig. 8.6 (a Charlesby–Pinner plot). There is a linear relationship between $S + S^{\frac{1}{2}}$, where S is the soluble fraction of the polymer, and the reciprocal of the dose for a polymer with a normal molecular weight distribution. The intercept at 1/dose = 0 is G (scission)/G (crosslink), and the gel dose corresponds to $S + S^{\frac{1}{2}} = 2$. Other distributions give curved plots, but the same intercept and gel dose.

Sometimes the energy required to form a crosslink or produce a permanent main-chain scission, indicated by E_c and E_s respectively, where E (in electron volts) = $100/G$, is used instead of the appropriate G value. G (crosslink) is usually greater at higher temperatures, e.g. for poly(ethylene) it has been observed to increase from 1 to 5 over the temperature range $-100°C$ to $+150°C$. The crosslinking yield for a particular polymer also depends on the physical state of the sample. Different values have been obtained for amorphous and crystalline bulk poly(ethylene) and for solution-crystallized poly(ethylene), where the polymer chains are folded in a regular pattern.

An interesting aspect of crosslinking in polymers is the memory effect. A polymer is crosslinked by irradiation and then heated and stretched. If cooled while stretched it retains this size or shape. However, any subsequent heating above the softening point causes shrinkage to the original dimensions. This technique has been used industrially for poly(ethylene) film and tube.

Other products

Hydrogen is the main gaseous product from the irradiation of poly(olefins) and also from many other polymers. This is partly due to the relative ease with which hydrogen atoms can diffuse away from a broken C–H bond and the rapidity with which H_2 is formed. An appreciable proportion of the hydrogen is also formed by molecular elimination, leaving vinylene (–CH=CH–) unsaturation. The trans form is predominant and can be measured from infrared bands. The cis form is not easily measured, but the total unsaturation can be determined, with varying degrees of success, by titration with bromine. For poly(ethylene) $G(H_2) = 3–4$ and $G(–CH=CH–) = 1–2$. The observation that frequently $G(H_2) = G(\text{crosslink}) + G(–CH=CH–)$ has been taken as evidence for reactions *8.18* and *8.19*. This relationship does not always hold and is not convincing.

$$-CH_2-CH_2- \rightsquigarrow -CH{=}CH- + H_2 \qquad (8.18)$$

$$2\,-CH_2-CH_2- \rightsquigarrow \begin{array}{c} -CH_2-CH- \\ | \\ -CH_2-CH- \end{array} + H_2 \qquad (8.19)$$

The yields for the main reactions in the radiolysis of poly(ethylene) are given in Table *8.2*.

The effect of radiation on insoluble polymers can be determined by measurements of bulk mechanical properties. These include tensile strength, elongation at break, the force required to maintain a constant elongation and hardness. These methods are particularly useful for polymers, such as rubbers, which are initially crosslinked. Radiation effects in industrial polymers are complicated by the presence of additives such as antioxidants and plasticizers.

Table 8.2. Radiation yields for radiolysis of poly(ethylene) in the solid state.

G(crosslink)	−100°C	1
	25°C	3
G(scission)		1
$G(H_2)$		3–4
G(−CH=CH−)		1–2

Polymers containing aromatic rings are relatively resistant to radiation, e.g. for polystyrene G(crosslink) ≈ 0.05. This is the same effect as in low-molecular-weight compounds. Some protection against radiation can be achieved in polymers by incorporation of aromatic additives, or by copolymerization with styrene.

The effect of radiation on polymers in solution may be quite different from the solid-state. In solution the polymer molecules provide only a small proportion of the total electron density. Therefore most of the primary radiation products result from the solvent, and these species attack the polymer. Hence indirect action is most important.

GRAFT COPOLYMERIZATION

Modification of the properties of a polymer by copolymerizing a second monomer on to it has become an important industrial technique. This is known as *graft copolymerization* and produces long side chains of the second polymer (B) attached to molecules of the first polymer (A).

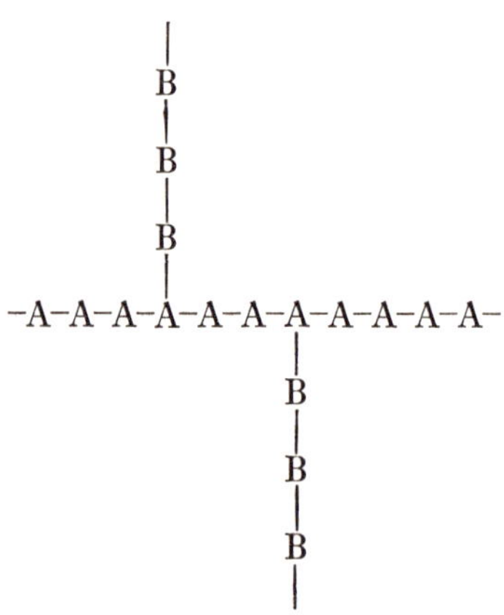

A micrograph of the cross-section of a typical graft copolymer is shown in Plate 8. If some of the polymerization of the monomer (B) is initiated independently of the substrate polymer (A), then polymer composed only of B, i.e. homopolymer of B, is also produced.

Grafting is particularly useful for changing the properties of the surface of a polymer without altering the physical and chemical characteristics of the bulk material. Films of non-polar polymers like poly(ethylene) and poly(propylene) can have monomers with polar substituents grafted on to them to enable printing. Synthetic fibres can be similarly treated so that they can be satisfactorily dyed. The permeability of plastic films to particular gases can be reduced by grafting on a monomer which produces an impermeable polymer layer.

The development of graft copolymerization has been largely for industrial uses, and much of the information is to be found in the patent literature. A variety of methods have been developed, but they all depend on the formation of an active site on the polymer chain. Irradiation of the polymer is a convenient method of achieving this. Four techniques have been used.

(1) The polymer is irradiated in the presence of the monomer, which may be present as vapour, liquid or in solution. Considerable amounts of homopolymer are usually produced. This method is suitable for monomers, such as styrene, for which G(initiation) in the monomer is low.
(2) The polymer is irradiated and then the monomer is added. This reduces homopolymer formation unless chain transfer to monomer is significant.
(3) The polymer is irradiated in air and then heated with the monomer. Peroxides are produced during irradiation, and these decompose on heating.
(4) Two polymers are irradiated in an intimate mixture. This produces crosslinking between the two types of chains.

Most studies of graft copolymerization have been with monomers which polymerize by a free radical mechanism. However, in principle, ionic grafting is quite feasible.

The amounts of grafted copolymer vary greatly, but 100% increase in weight can be obtained. Swelling of the substrate greatly increases the amount of grafting. Increases in grafting at high temperatures are usually attributed to greater swelling. If a monomer–polymer mixture is irradiated and radicals are readily produced in the graft chains, then autoacceleration may occur.

Grafting on to cellulose has been widely studied and is very efficient with swelling agents. Commercial applications are developing and may become large-scale processes. Production of crease-resistant clothing by graft copolymerization on to cotton or wool is one important field. Another is the impregnation of wood with monomer, followed by radiation-induced polymerization. The proportion of monomer may be large, and then homopolymerization is probably more important than grafting to the cell walls.

An interesting application of graft copolymerization is the attachment of chemically reactive groups to an insoluble polymer, e.g. poly(tetrafluoroethylene) or crosslinked poly(styrene). One example is the assay of antibody by reaction with antigen coupled to an insoluble graft copolymer of styrene on poly(tetrafluoroethylene). Another example is the preparation of polypeptides by progressive condensation of amino acids on to cross-linked poly(styrene).

BIOLOGICAL MACROMOLECULES

The effects of radiation on *biological macromolecules*, such as proteins, can be classified into those in aqueous solution and those in the solid-state. Indirect attack by hydroxyl radicals, hydrogen peroxide and hydrated electrons are the most important reactions in aqueous solution. There is evidence of L.E.T. effects. Irradiation in the solid-state produces direct cleavage of bonds. As in all polymers, permanent breakage of bonds to hydrogen occurs, due to the relative ease with which the hydrogen atom can escape from the cage. This results in disruption of the 'tertiary' structure of biological macromolecules, which is strongly dependent on hydrogen bonding. Consequently, the macromolecules become increasingly susceptible to chemical attack, especially at disulphide linkages, leading to further breakdown. Biological macromolecules are damaged by extremely small doses of radiation, and this seems to result from high specificity for certain sites. Migration of energy or charge along the chain to a reactive site may be an important factor in this.

Electron-spin-resonance studies have shown that free radicals are produced in irradiated biological macromolecules in the solid-state. The lines of the spectra are usually broad and overlap. However, the asymmetry of the spectra suggest that the unpaired electron is frequently located near a sulphur atom. These free radicals have long lifetimes, even at room temperature, and provide sites for grafting. Protection against radiation damage can be achieved by addition of –SH compounds, and this is apparently due to radical scavenging.

More information on biological systems will be found in Chapter 9.

FURTHER READING

BATTAERD, H. A. J. and TREGEAR, G. W., *Graft Copolymers*, Wiley, New York (1967).

CHAPIRO, A., *Radiation Chemistry of Polymeric Systems*, Interscience, New York and London (1962).

CHARLESBY, A., *Atomic Radiation and Polymers*, Pergamon, Oxford (1960).
HENGLEIN, A. and SCHNABEL, W., Radiation Chemistry of Synthetic Macromolecules in Solution, L.E.T. Effects, in *Current Topics in Radiation Research*, EBERT, M. and HOWARD, A. (Eds.), North Holland, Amsterdam, Vol. 2, p. 1 (1966).
Irradiation of Polymers, *Advances in Chemistry Series*, No. 66, American Chemical Society (1967).
OSHIMA, K. and TABATA, Y., Radiation-Induced Polymerization at Low Temperatures, in *Advances in Nuclear Science and Technology*, Vol. 3, Academic, New York (1966).
† SCHULTZ, A. R., Cleavage Reactions—Radiation and Crosslinking by Radiation, in *Chemical Reactions of Polymers*, FETTES, E. M. (Ed.), Interscience, New York (1964), pp. 645 and 723.

† These references should be read first.

Chapter 9

Radiation Biology and Biochemicals

Although radiation has been of considerable use in medical diagnosis and therapy, the hazards of radiation have also attracted much attention. The hazards to life are of two principal kinds: acute injury, which is evident soon after exposure to radiation, and delayed reactions, which include malignant disease and genetic abnormalities.

Biological material is immensely complex in chemical terms, and its make-up is far from being understood. Since identification and interpretation of the reactions involved in the radiolysis of even the simplest chemical systems are extremely difficult, extension to biological systems presents a formidable challenge. Nevertheless, considerable progress has been made and, in the process, much useful information on biological structure and function has been discovered.

The reader should refer to appropriate texts, reviews and research papers for more information on this progress. We will merely summarize some of the main features before dealing with the radiation chemistry on which it is based. Radiation biology should be investigated with the aim of relating the biological changes to the chemical reactions produced by the radiation. Experiments with whole animals or cells may indicate which physiological structures or biochemical processes merit further attention at a chemical level. Conversely, radiation chemistry studies of bio-chemical compounds such as amino acids may indicate the manner in which proteins or nucleic acids behave in irradiated organisms. Both lines of investigation are necessary if we are to understand the biological actions of radiation.

RADIATION DAMAGE

Biological radiation damage is essentially cellular. Therefore, its causes must be sought within the individual cells rather than extra-cellular supporting structures or nutrient fluids, although the changes induced in these by radiation may be of chemical interest.

Very low radiation doses, of the order of 500 rad (5 mJ/g), are needed to destroy the dividing cells of higher organisms and, in consequence, the organisms themselves. Simple, unicellular organisms, such as bacteria, may be more resistant, but the radiation sensitivity of bacteria varies enormously. Also, the doses required to produce delayed effects may be considerably less than those which produce acute effects.

All the coded information needed to define a cell's growth, structure and function is contained in the deoxyribonucleic acid (DNA) of its chromosomes. The DNA of a cell is paired, one half being traceable to the paternal, and the other half to the maternal, parent. It is the only component of the cell which is not freely replaceable. Any change to the DNA which will lead to faulty reduplication of DNA when the cell divides must be a potential cause of cellular malfunction or inherited abnormality. Therefore, radiobiology is very much concerned with the radiation chemistry of DNA and its constituents.

However, the radiation chemist seeking to explain radiation biological effects must go further than DNA. The effects of high doses on non-dividing cells may be due predominantly to injury to cellular membranes. The radiation chemistry we are concerned with in this case is that of protein and phospholipid monolayers.

Inactivation of cells

A criterion of radiation damage that is often used in investigations of bacteriophages, bacteria and mammalian cells is inactivation or reproductive death. If we irradiate any of these and plot the number of survivors against the radiation dose, the resulting curve will usually

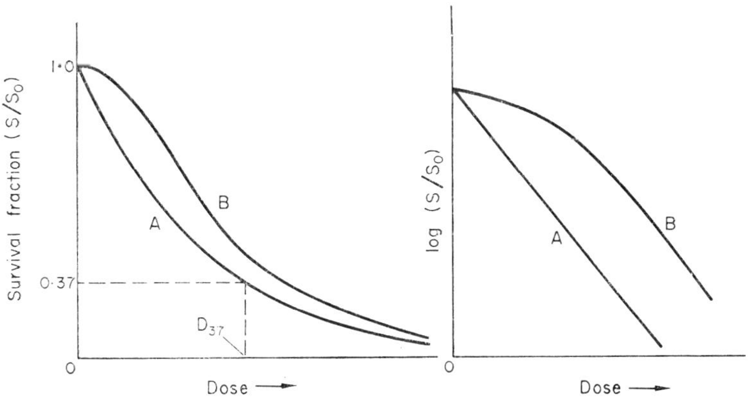

Fig. 9.1 Inactivation. Survival fraction (S/So) and log (S/So) plotted against dose. (A) exponential, (B) sigmoidal relationship.

approximate to either an exponential or a sigmoidal relationship. These results are usually plotted as the logarithm of the survival fraction versus dose as shown in Fig. 9.1.

The simple exponential relationship would be expected if a cell (or other unit) is inactivated by a single harmful event or 'hit', and if the probability of a hit is proportional to the dose. The sigmoidal relationship results from a more complex model. This assumes that for a cell to be inactivated a number of 'targets' within the cell must be inactivated and that each target is inactivated by a single hit. A sigmoidal curve is also obtained when there is a distribution of radiation resistance about a mean value. For the sigmoidal relationship there is an apparent *threshold dose*, below which no radiobiological effect is observed. In exponential systems a convenient measure of the radiation sensitivity is the lethal dose, corresponding to an average of one lethal hit per target, i.e. per individual. This will give 37% survival, since statistically this fraction of targets will escape, and is often designated by the symbol D_{37}.

Repair phenomena

It has long been known that there is complete or partial recovery from acute radiation damage in a whole animal. This phenomenon can be interpreted in terms of restoration of the cellular systems injured by the radiation and an absence of residual damage in the recovered cells, other than the potential of delayed effects. It is found that if a certain dose is given in two parts, the survival fraction is greater than that following a single exposure. If the interval between the two doses is progressively increased it is found that the cells which survive the first dose eventually behave as if complete recovery had taken place after the first exposure.

Repair of specific chemical damage to DNA in bacteriophages and bacteria caused by ultraviolet light has been observed. It may occur by photo-reactivation (in visible light), or by 'host-cell reactivation'. Similar phenomena may occur in the cells of higher organisms and with respect to damage caused by ionizing radiation. Further progress in the understanding of biological radiation injuries may well result from identification of specific chemical damage and studies of repair mechanisms.

Radiosensitivity

It is possible to modify the radiation sensitivity of whole organisms or of cells by various means, but in general these will not alter the mean lethal dose by more than a factor of 2 or 3 in either direction. The best known of these means is oxygen: the sensitivity of cells is 2–3 times greater in air than in an anoxic state. However, the details of the mechanism remain obscure. Other agents which may sensitize cells to radiation include metabolic inhibitors, which may hamper repair processes, and base analogues such as bromouracil, which can partly replace thymine (methyl uracil) in the DNA molecule and thereby increase its radiosensitivity. There are also a large number of types of compound which decrease radiosensitivity, but, with the exception of those which clearly act by producing local anoxia, their modes of action are also rather obscure. They must all be present at the time of irradiation to be effective and are rather too toxic to have much practical application.

L.E.T. and R.B.E.

A factor which must be considered in any evaluation of the primary effects of radiation is linear energy transfer (L.E.T.). In general, the relative biological effectiveness (R.B.E.) of radiation (relative to that of 250-kV x-rays) increases with L.E.T. and for some particulate radiations may be as high as 10–20. With further increase the R.B.E. may fall. The actual values of R.B.E. found depend very much on the biological end point or test chosen. It is not difficult to see that a densely ionizing track may be very effective in producing a chromosome injury, where several ionizing events may be required, but relatively inefficient in damaging a bacteriophage, where only one event is necessary.

BIOCHEMICALS

The effects of radiation on the chemical components of biological systems and on simpler model compounds have been studied extensively. Both direct and indirect effects are important (see Chapter 2). When the radiation acts indirectly, the radiation effect produced by a given dose is independent of the concentration of the solute or organism. This can be readily distinguished from direct action, for which the effect increases proportionately with concentration as shown in Fig. 9.2.

Aqueous solution chemistry is particularly relevant to radiation biology. In Chapter 6 we discussed the reactive species formed in the radiolysis of water–hydrated electrons, hydrogen atoms, hydroxyl radicals and, in the presence of dissolved oxygen, hydroperoxyl radical ions. The reactions of these species with various biochemical compounds have been investigated in detail. Many of the more radiation-sensitive biological compounds, such as DNA, are macromolecules, and the principles discussed in Chapter 8 also apply to them.

There is no doubt that electrons play an important role in many biological processes, although in some reactions the part played by hydrogen atoms appears to be established. It is interesting to note regarding the anions of the Krebs cycle, a metabolic process occurring in the mitochondria of biological cells, that alternate members react readily or poorly with hydrated electrons. Thus, pyruvate, α-ketoglutarate, fumarate and oxalacetate react readily, while citrate, isocitrate, succinate and malate do not react readily.

We will summarize the main features of the radiation chemistry of some types of compounds which are important in biological systems.

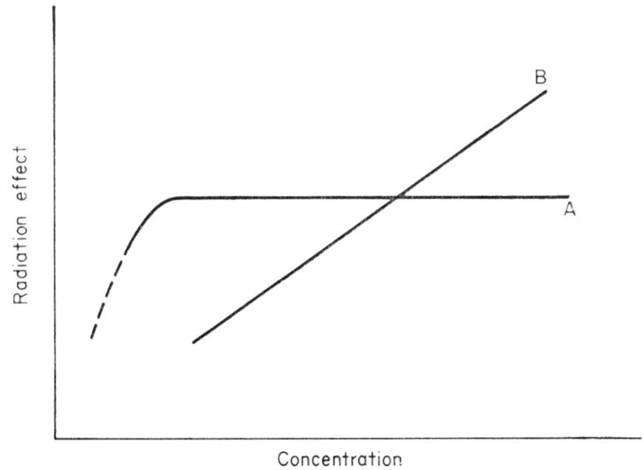

Fig. 9.2 Distinguishing between (A) indirect and (B) direct effect of radiation.

Carbohydrates and polysaccharides

Carbohydrates and polysaccharides are important components of plants and animals. Sucrose is a dimer, and starch, cellulose, dextran and glycogen are long-chain polymers of hexose sugars (or monosaccharides) of different structure or with different linkages.

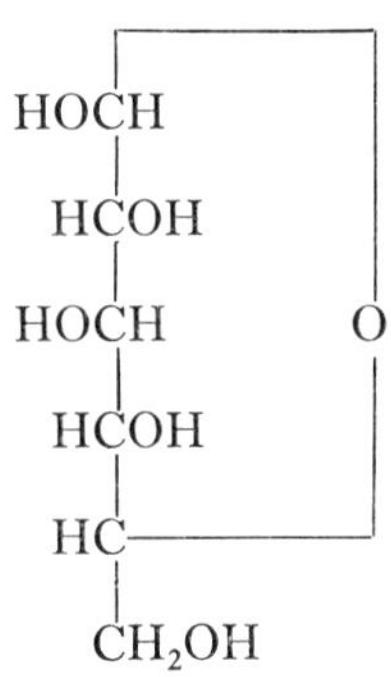

glucose

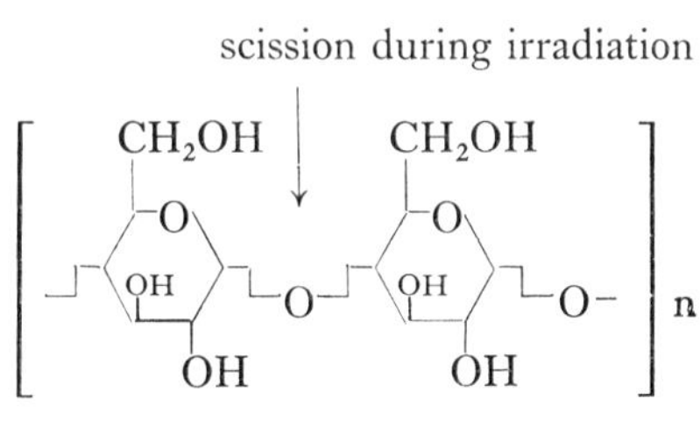

glycogen

The polymers undergo rapid degradation in the solid-state or in solution. Random scission of the backbone chain occurs with G values as high as 13, which appears to result from specific breakdown of the relatively weak glycoside inter-unit linkages. There is some associated production of monomer units, e.g. glucose, and similar species, such as gluconic acid. The solution properties of the polymers, e.g. viscosity and light scattering, suggest that some chain branching may also occur. Electron-spin-resonance spectra show that free radicals are trapped in the solid, and these will react slowly on standing, or rapidly on dissolution.

Degradation of sucrose (a disaccharide) to the monosaccharides, glucose and fructose, i.e. inversion, is brought about by irradiation. Radiolysis of monosaccharides, such as glucose, results in some formation of dimeric and higher polymers (the equivalent of branching in the radiolysis of polysaccharides).

A variety of chemical reactions also take place at other sites in the mono- and polysaccharides, particularly at the hydroxyl groups.

Hydroxyl radicals and hydrogen atoms abstract hydrogen from simple sugars, whereas hydrated electrons are relatively unreactive.

Proteins

There has been a considerable amount of work done on the effect of radiation on proteins and on their basic units, the α-amino acids.

The zwitterion forms of α-amino acids react fairly readily with hydrated electrons, whereas peptides are rather more reactive. In both cases the activity is located in the protonated amino group ($-NH_3^+$). Amino acids, peptides and proteins containing reactive side-chain groups (e.g. –S–, rings) are very reactive in all ionic forms with hydrated electrons, hydrogen atoms and hydroxyl radicals.

The proteins are copolymers containing specific sequences of monomers selected from a group of twenty α-amino acids. They are a class of polypeptide, as they contain the peptide inter-unit linkage, $-\overset{\overset{O}{\|}}{C}-\overset{\overset{H}{|}}{N}-$.

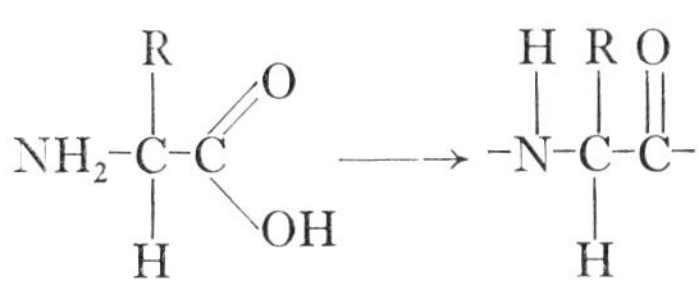

(9.1)

α-amino acid polypeptide

Irradiation of amino acids produces CO_2 and the corresponding amine. Scission of C–H bonds and dimerization occurs. In aqueous solution NH_3 is frequently produced.

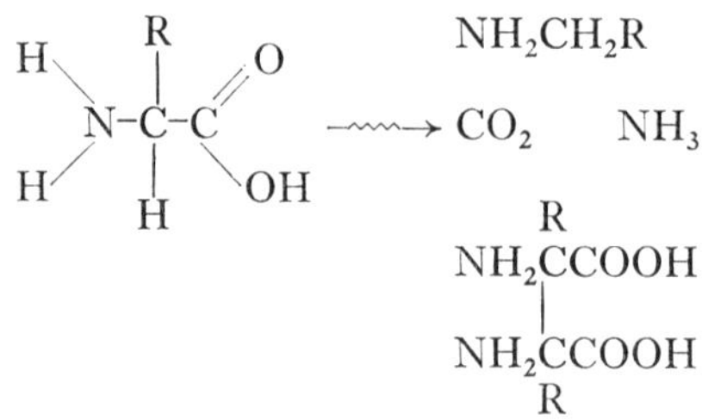

(9.2)

When the amino acid contains an aromatic ring, the effect of the irradiation is reduced, as expected, and in aqueous solution some hydroxylation of the aromatic ring occurs.

When proteins, or synthetic polypeptides, are irradiated the peptide inter-unit linkage is susceptible to attack, particularly in aqueous solution, and NH_3 is produced. Amino-acid residues containing sulphur, such as cysteine, are readily attacked by free radicals in solution or in the solid, except at low temperatures. Small amounts of H_2S may be produced, probably by direct radiolytic scission of C–S bonds, followed by combination of the HS fragment with a hydrogen atom.

The radiolysis of simple peptides in very dilute aqueous solution has been interpreted in terms of hydrogen-atom abstraction from the peptide by hydroxyl radicals,

$$\cdot OH + RH \longrightarrow H_2O + R\cdot \qquad (9.3)$$

followed by dimerization.

$$R\cdot + R\cdot \longrightarrow RR \qquad (9.4)$$

Studies have been made on the e.s.r. spectra of irradiated polypeptides, including single crystal studies on dipeptides. The spectra have been attributed to alkyl radicals, and in sulphur-containing compounds to $RCH_2S\cdot$.

The principal biological interest in the radiation chemistry of proteins has been concerned with the effect on enzymes. The decrease in enzymatic activity is usually exponential with dose (Fig. 9.3), and increasing radiation sensitivity with increasing temperature has been widely observed.

The formation, transformation and disappearance of radicals in irradiated enzymes has been studied by e.s.r. The identity of the primary radicals formed during irradiation at $-196°C$ is uncertain, as the spectrum is a very broad singlet over a range of 70 gauss about the free spin g value. Secondary radicals are formed on warming, or irradiation at higher temperatures, and are attributed to sulphuryl radicals, probably produced by the reaction.

$$R\cdot + R'SH \longrightarrow RH + R'S\cdot \qquad (9.5)$$

A correlation between radical formation and enzyme inactivation is frequently observed, but probably one does not result from the other, but rather they both indicate the amount of damage.

An early theory suggested that a single ionization anywhere in an enzyme molecule would cause inactivation. Excitations in the enzyme or ionizations outside the molecule were believed to have little effect. This was supported by the observation that radiation sensitivity apparently increases with the size of the enzyme molecule. However, recent work

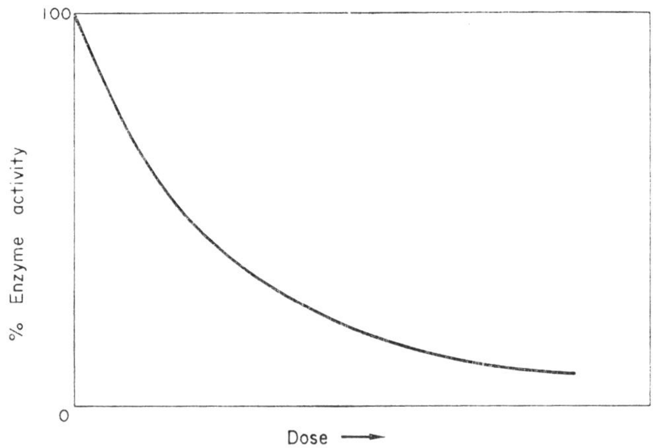

Fig. 9.3 Effect of radiation on enzyme activity.

indicates that many ionizations may occur in each molecule before de-activation.

Nucleic acids

There has been a considerable amount of work done on the nucleic acid bases. Both pyrimidine and purine bases react at the double bonds with hydrated electrons, hydrogen atoms and hydroxyl radicals. The effect of radiation on nucleic acids is best understood by considering their structure. Every nucleic acid is formed from a series of nucleotides. Each nucleotide consists of three components: (1) a sugar—either D-ribose in RNA, or D-2-deoxyribose in DNA, (2) a base and (3) phosphoric acid. Each DNA 'molecule' is in the form of a double-strand helix with an invariant pairing of opposite bases. In complex living material it is not really possible to define a molecular weight for DNA; it may even extend to the length of the chromosome.

Radiation damage to nucleic acids may be major—chain scission or crosslinking to the complementary strand, another nucleic acid molecule or other cell components—or minor changes to individual bases. It is accepted that all chemical changes in DNA do not lead to loss of reproductive power. Also, post-irradiation repair processes may eliminate damage to the double-stranded helix conformation.

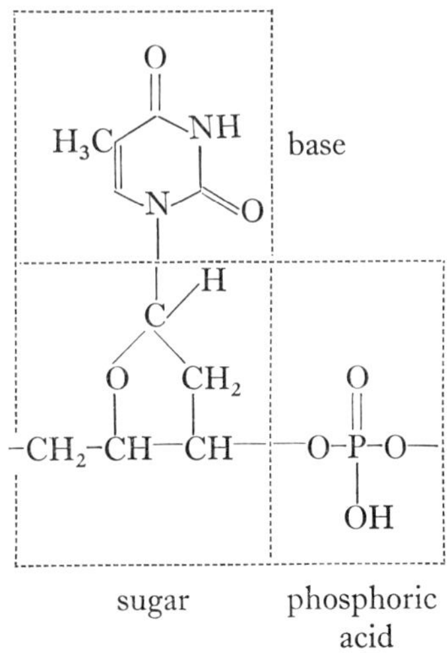

Structure of a nucleotide in DNA

Larger doses of radiation produce gross effects. The double-strand helix is destroyed and considerable chain degradation occurs. This probably proceeds by scission of the phosphate linkage, and phosphate esters and inorganic phosphates are produced. A variety of reactions also occur in the bases and the sugar.

Lipids

Animal fat consists of glycerides of saturated and unsaturated fatty acids from C_{12} to C_{20}.

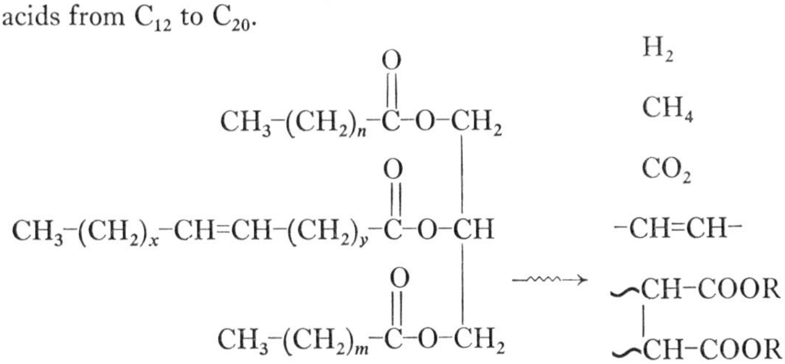

(9.6)

Typical lipid: a triglyceride of fatty acids.

The effects of radiation are those to be expected for hydrocarbon chains and carboxyl groups. Hydrogen is a major product, unsaturation is produced and CO_2 may be evolved. Some chain scission occurs and crosslinking, mainly via the carbon atoms α to the carboxyl groups, takes place. Reactions involving hydroperoxides may be important.

FURTHER READING

† ADAMS, G. E., The General Application of Pulse Radiolysis to Current Problems in Radiobiology, in *Current Topics in Radiation Research*, EBERT, M. and HOWARD, A. (Eds.), North Holland, Amsterdam, Vol. **3** (1967), p. 35.

Advances in Radiation Biology Series, Academic, New York (1964 onwards).

ALEXANDER, P. and LETT, J. T., Effects of Ionizing Radiation on Biological Macromolecules, in *Comprehensive Biochemistry*, FLORKIN, M. and STOTZ, E. H. (Eds.), Elsevier, Amsterdam, Vol. **27** (1967), p. 267.

BACQ, Z. M., *Chemical Protection Against Ionizing Radiation*, Thomas, Springfield, Illinois (1965).

BACQ, Z. M. and ALEXANDER, P., *Fundamentals of Radiation Biology*, Pergamon, Oxford, 2nd edn. (1961).

CASARETT, A., *Radiation Biology*, Prentice Hall, New Jersey (1968).

GINOZA, W., The Effect of Radiation on Nucleic Acids of Bacteriophages and Bacterial Cells, *Ann. Rev. Nucl. Sci.*, Vol. **17**, 469 (1967).

LEA, D. E., *Actions of Radiations on Living Cells*, Cambridge University Press (1956).

MORGAN, K. Z. and TURNER, J. E., *Principles of Radiation Protection*, Wiley, New York (1967).

ROBERTSON, J. S. and BOND, V. (Eds.), *Studies in Radiation Effects*, Series B: Biological and Medical, Gordon and Breach, New York.

† WHITFIELD, R. E. and WASLEY, W. L., Effects of High Energy Radiation on Proteins, in *Chemical Reactions of Polymers*, FETTES, E. M. (Ed.), Interscience (1964).

Workshop Conference on Space Radiation Biology, *Radiation Research* Supplement 7 (1967).

ZIMMER, K. G. and MULLER, A., New Light on Radiation Biology from Electron Spin Resonance Studies, in *Current Topics in Radiation Research*, EBERT, M. and HOWARD, A. (Eds.), North Holland, Amsterdam, Vol. **1**, (1965) p. 1.

† These references should be read first.

Chapter 10

Some Other Systems

The effects of radiation on a wide variety of systems have been studied for reasons ranging from obtaining a better understanding of principles to the development of industrial processes. We have selected a few systems for discussion to illustrate the wide ramifications of radiation chemistry.

INORGANIC SOLIDS

The predominant effect of radiation on inorganic solids is to displace atoms or ions, producing defects in the lattice structure. These defects are responsible for colouration of the sample and for changes in physical properties, particularly electrical conduction. A *lattice vacancy* will be produced when an atom or ion is displaced and an *interstitial defect* formed

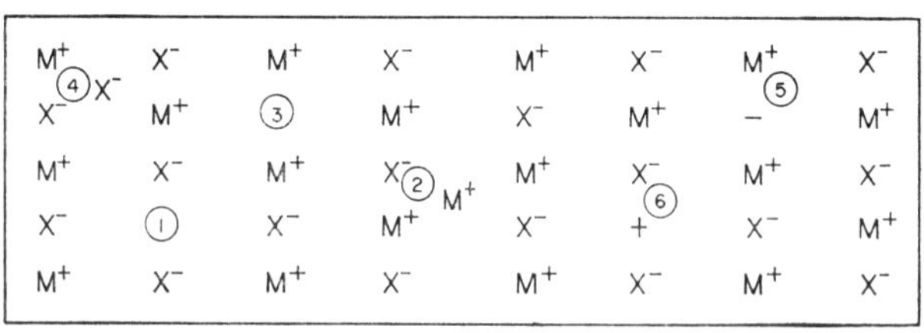

Fig. 10.1 Schematic diagram showing defects produced in inorganic solids by irradiation. (1) cation vacancy, (2) cation interstitial, (3) anion vacancy, (4) anion interstitial, (5) F-centre, (6) V-centre.

if the atom or ion is trapped between the lattice planes (Fig. 10.1). Obviously, vacancies and interstitial defects will be formed in equal numbers unless other reactions occur. The total energy requirement for a vacancy plus an interstitial species is about 25 eV (4 aJ). Although only 1 eV (0.16 aJ) is required to actually form the vacancy and 4 eV (0.64 aJ) for the interstitial species another 20 eV (3.2 aJ) is required for the disruption of the neighbours.

These defects may be produced by *direct displacement* (knock-on) or as a result of ionization. Knock-on occurs mainly with neutron irradiation, and each fast neutron may cause a cascade of up to 10^3 successive displacements. Momentum transfer by β- or γ-radiation is relatively small and generation of defects occurs mainly through ionization.

The effect of radiation on inorganic solids is becoming increasingly important with the widespread use of semiconductor devices in radiation fields and in extra-terrestrial communications.

Electron ejection is of no importance in metals because of the existing

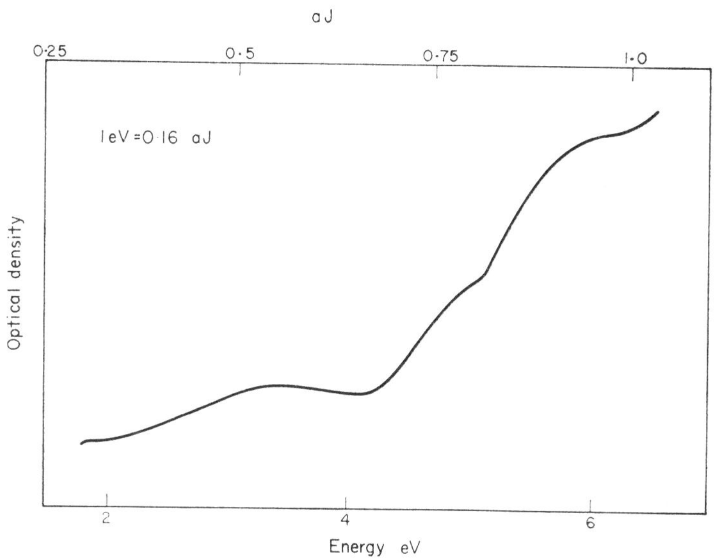

Fig. 10.2 Optical absorption in Al_2O_3 due to saturation yield of colour centres produced by γ-irradiation.

freely circulating electron cloud. However, in non-metals it can cause marked changes in properties. An electron can be readily trapped in an anion vacancy; this is known as an *F-centre*. It causes colouration and can be detected by e.s.r. The e.s.r. spectra can provide information on the interaction of the trapped electron with the neighbouring ions, and frequently several types of F-centre can be distinguished. A positive charge can be trapped in a cation vacancy, and this is known as a *V-centre*. It can also be detected by e.s.r. Formation of an F or V centre necessitates a change in valence of an ion.

Colour centres and other radiation-induced defects in inorganic solids may be removed by heat treatment or by ultraviolet light. After irradiation at low temperatures, e.g. 4°K or 77°K, considerable annealing takes place with only a small temperature rise. At certain temperatures, corresponding to defect or trap energies, strong luminescence is frequently observed. Impurities, which cause discontinuities in the lattice structure, play a major role in annealing processes. Similarly, other lattice defects, such as slip planes, are important. Radiation annealing also occurs, particularly for F- and V-centres, as a result of the continuous production of mobile electrons by ionization. Consequently, limiting concentrations of defects are frequently formed after moderate radiation doses. Fig. 10.2 shows the absorption spectrum of Al_2O_3 due to the equilibrium or saturation concentration of colour centres produced by γ-radiation (saturation dose 0.5 Mrad).

Chemical reactions also occur in inorganic solids during irradiation. For example, nitrates decompose to give nitrites and oxygen.

$$NO_3^- \rightsquigarrow NO_2^- + \tfrac{1}{2}O_2 \qquad (10.1)$$

The radiation chemical yield (G value) varies with the metal ion. It is most unlikely to depend on the chemical properties of the metal ion; rather it must be related to the physical parameters of the lattice, e.g. the closeness of packing. This is a good example of a *topochemical* effect. Electron-microscope studies of reactions of this type indicate that disruption of the lattice occurs in localized regions, which then grow in size. This suggests that reaction proceeds predominantly where the crystalline structure has been broken down.

Chlorates

The inorganic chlorates provide a good example of the complexity of products which can be expected when oxidation and reduction reactions are possible. Table 10.1 shows the yields obtained from the γ-radiolysis of solid potassium chlorate.

Table 10.1. Radiolysis products from solid $KClO_3$

Product	G value
O_2	2.8
ClO_2^-	2.1
Cl^-	1.6
ClO_4^-	1.1
ClO^-	0.3
ClO_2	0.2
H_2	0.1

The products are separated by dissolving the irradiated solid in water, and this explains the hydrogen yield, which is attributed to the reaction of electrons, trapped in the solid, with the water.

$$2e^-_{trap} + 2H_2O \longrightarrow H_2 + 2OH^- \qquad (10.2)$$

Decomposition of excited chlorate ions is believed to give O_2, O, ClO_2^-, Cl^-, ClO^- and ClO_2 as primary products, by reactions of the type,

$$ClO_3^- \rightsquigarrow ClO_2 + O + e^- \qquad (10.3)$$

$$ClO_3^- \rightsquigarrow Cl^- + O_2 + O \qquad (10.4)$$

The ClO_4^- may be formed by oxidation of ClO_3^- by oxygen atoms,

$$ClO_3^- + O \longrightarrow ClO_4^- \qquad (10.5)$$

or it may be produced by decomposition of Cl_2O_6 on dissolution.

$$2ClO_3 \longrightarrow Cl_2O_6 \xrightarrow{H_2O} ClO_4^- + ClO_3^- + 2H^+ \qquad (10.6)$$

A proportion of the oxygen will result from combination of oxygen atoms.

$$O + O \longrightarrow O_2 \qquad (10.7)$$

Similar relative yields of the different products have been obtained from different chlorates, indicating that the cation does not play an important role in determining the mode of decomposition. However, the radiation sensitivity varies, and this has sometimes been correlated with a concept of 'free space' within the crystal.

perbromate has been detected in irradiated bromate. This is interesting because perbromate is the missing perhalate.

Quartz, silica and glass

The silicon oxide structures provide a complex, but interesting, class of compounds for radiolysis studies. Although they all consist basically of –Si–O– networks, they show a great variation in properties and radiation sensitivity. The physical state may be crystalline (quartz), or vitreous, and there may be a narrow or a broad melting-point range. Impurity metal ions have a major effect on the physical properties and on the radiolysis.

There are two reasons for the interest in the effects of radiation on glass. Firstly, the effect of radiation is in itself important because glass is such a widely used material of construction. Secondly, it is a useful method for investigating the structure of glass.

Heavy particle radiation can produce structural changes, but electrons and γ-rays cause only electron ejection and trapping. These electronic effects are associated with existing structural defects, which are usually the result of impurity atoms.

Charge trapping may be observed visually, or by e.s.r. Thus, Pyrex glass turns brown after small doses of radiation. Irradiated wine glasses have a pleasant, smoky-grey appearance. Higher doses (10^5 rad) are required to produce visible colouration (violet) in ordinary silica and 10^7 rad is necessary for high-purity silica. A typical absorption spectrum of irradiated silica is shown in Fig. 10.3; three different types of colour centre are evident.

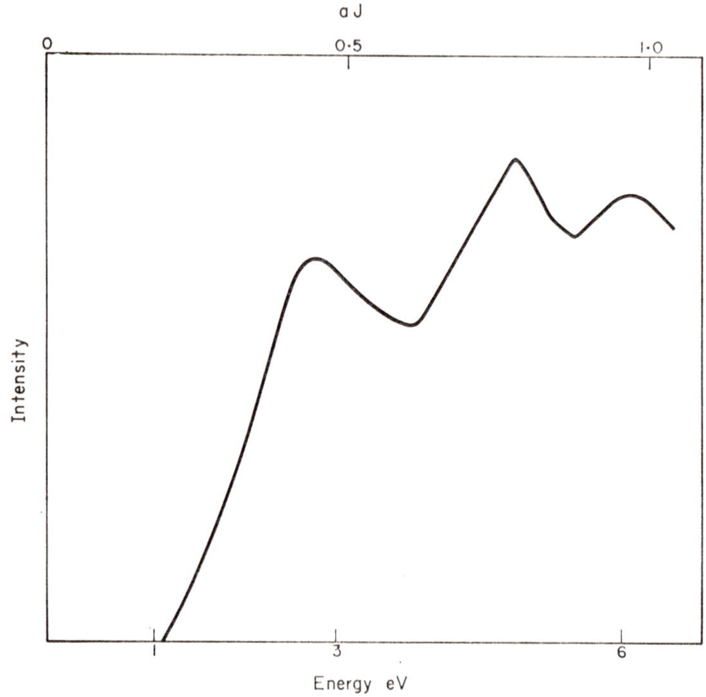

Fig. 10.3 Absorption spectrum of γ-irradiated silica.

Small amounts of deliberately added metal ions can have marked effects. For example, cerium acts as a protective agent against radiation-induced colouration. This can be compared with the use of manganese to protect window glass against colouration by sunlight. The additive acts by changing the type, rather than the number, of colour centres. As in other systems, the colour centres are frequently bleached by ultraviolet light, and also by warming.

A feature of the irradiation of some glasses at high dose rates of electron or γ-ray beams is the accumulation of charge. An electrical discharge can occur and rupture the glass.

HETEROGENEOUS CATALYSIS

Heterogeneous catalytic effects in radiation chemistry can be divided into two categories, depending on whether (1) the substrate and reactants are irradiated together, or (2) the solid catalyst is irradiated separately.

(1) Solids added to a liquid- or gas-phase system sometimes affect the rate of reaction occurring in that phase during irradiation. Thus, solid zinc oxide increases the oxidation of benzene in aqueous solution from $G = 6$ to 16.

The added solid may act in a number of ways. Molecules (M) absorbed on the surface of the solid may react with F-centres (electrons trapped in anion vacancies) to form anions, or with V-centres (positive holes trapped in cation vacancies) to form cations.

$$M + (-) \longrightarrow M^- \qquad (10.8)$$

$$M + (+) \longrightarrow M^+ \qquad (10.9)$$

Subsequent neutralization of the anion or cation with a V-centre or an F-centre respectively may produce sufficient excitation energy for dissociation into radicals to occur.

$$M^- + (+) \longrightarrow M^* \longrightarrow R_1\cdot + R_2\cdot \qquad (10.10)$$

$$M^+ + (-) \longrightarrow M^* \longrightarrow R_1\cdot + R_2\cdot \qquad (10.11)$$

Alternatively, charge transfer or ionic reactions, such as polymerization, can occur.

$$M^+ + N \longrightarrow M + N^+ \qquad (10.12)$$

$$M^+ + M \longrightarrow M_2^+ \qquad (10.13)$$

Dissociation is favoured when the excess energy is at least 4 eV (0.64 aJ), otherwise electron transfer may occur. These catalytic effects occur only with the first absorbed layer, and only a fraction of this layer is actually active. The fall in catalytic activity with surface coverage is shown in Fig. 10.4. Increases in the overall rates of chain reactions, such as addition polymerization, of up to 10 times have been observed. These can only be accounted for if the chain termination rate is decreased. This is attributed to decreased mobility of the chain radicals, hence they are apparently adsorbed on the surface. Confirmation of this has been provided by the increased molecular weight of polymers.

(2) Irradiation of a solid catalyst sometimes alters its efficiency for catalysing a particular chemical reaction. This effect may be evident (*a*) during irradiation, or (*b*) after irradiation.

Increased catalytic activity requires that irradiation should increase the number of catalytically active sites on the surface of the catalyst. These sites are usually structural defects, hence irradiation is frequently effective.

A recent technique of both academic and practical interest is the incorporation of a radioactive isotope into the catalyst. This internal irradiation should be highly effective in producing defects continuously in the lattice of the catalyst and in promoting electron exchange reactions with adsorbed species.

The increases in reaction rates which are possible by radiation catalysis are of considerable industrial interest, and a number of processes are being investigated.

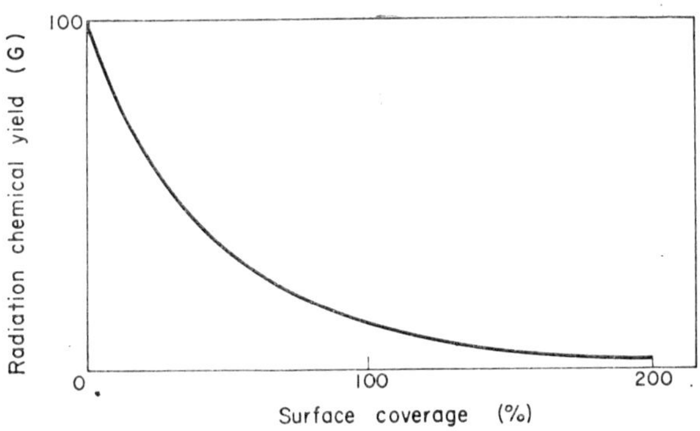

Fig. 10.4 Variation in radiation chemical yield (G value) with surface coverage of solid catalyst.

Electrode processes

Electrodes incorporating radioactive materials have been found to have lower over-potentials than inactive materials. This has been suggested as a means of increasing the efficiency of fuel cells. Radiation increases the rate of corrosion of the components of nuclear reactors. Similar considerations may be involved here.

HOT ATOM CHEMISTRY

In Chapter 3 it was pointed out that following neutron capture by some elements the immediate daughter product, i.e. the composite or compound nucleus, was radioactive. Very soon after neutron capture the composite nucleus eliminates excess energy as one or more γ quanta.

$$^{35}_{17}\text{Cl} + ^{1}_{0}\text{n} \longrightarrow ^{36}_{17}\text{Cl} + \gamma \qquad (10.14)$$

To conserve momentum the daughter nucleus recoils, and often this recoil is energetic enough to break chemical bonds.

The recoiling nucleus (and attendant electrons) loses energy by interacting with the medium, becoming neutralized in the process. Finally, at an energy less than 100 eV (16 aJ), it is able to undergo *hot reactions*, The study of these reactions is called *hot atom chemistry*.

During the slowing down of the charged, recoiling atom many molecules of the medium might be ionized or excited, and therefore radiation chemical effects will be produced. By the use of appropriate scavenging and other techniques these may be distinguished from hot atom reactions.

MATERIALS

Some materials can be used in high radiation fields and suffer very little damage, but others are quite sensitive to radiation. Heavy particles (fission fragments, α-particles, protons and neutrons), even of relatively low energies, are capable of displacing atoms from a crystal lattice, and some radiation damage is then manifest in almost all materials. For instance, the surface of a glass vessel in contact with a solution containing an α-emitter becomes crazed with small cracks. Light particles (β-particles and electron beams) and photons (γ-rays and x-rays), on the other hand, require energies of some hundreds of thousands of electron volts to displace atoms as heavy as oxygen from crystal lattices. Most of their effects result from ionization and excitation and the subsequent chemical reactions of the species resulting from these processes.

Metals are very resistant to light-particle and photon radiation, as are ceramics, mineral materials and concrete. Glasses and solid alkali halides become coloured due to the development of F-centres, but retain their physical properties. The colour can be annealed out of them. Ordinary silica becomes a violet colour under irradiation, whereas the purest forms of quartz give hardly any colour at all, and may be used as optical windows or in e.s.r. studies.

Wood is reasonably resistant to radiation. After about 50 Mrad of irradiation, plywood is shattered by a light blow. Cork is somewhat more resistant than wood.

Lubricating oils containing aromatic rings are less sensitive to radiation than are long-chain oils. Inorganic lubricants, such as graphite and molybdenum disulphide, are very resistant.

Coolants for use in a radiation field or in a nuclear reactor must be selected for radiation resistance. Liquid metals are particularly resistant. Polyphenyls are resistant enough to be used in organic moderated and cooled reactors and have been extensively studied. Water is satisfactory for many purposes. If its purity can be maintained at a high level, there is little decomposition over a long time in a high-radiation field.

Plastics might be used in a radiation field as electrical insulation, as gaskets and seals, structurally or for wrapping. Generally the desirable properties are improved at low doses, but deteriorate at higher doses. The deterioration is more rapid when oxygen is present.

The following generalizations can be made regarding the stability of plastics. Poly(urethanes), poly(styrene) and epoxy resins are the most stable. Phenolic resins, poly(esters) with styrene, poly(ethylene), poly-(ethylene terephthalate), rubber, neoprene and nylon are satisfactory up to 100 Mrad. Poly(vinyl chloride), urea–formaldehyde resins, cellulose acetate and most textiles are satisfactory up to 10 Mrad and poly(methyl methacrylate), poly(propylene) and cellulose up to 2.5 Mrad. Poly(tetra-fluorethylene) and casein are generally unsatisfactory. Chlorinated plastics such as poly(vinyl chloride) and neoprene usually liberate hydrogen chloride on irradiation, and this may cause corrosion or other undesirable effects.

ADDITIONAL SYSTEMS

A number of rather unusual systems have been investigated in the hope of providing a better insight into what happens on irradiation, or of revealing possible applications for radiation in industry. These include liquid ammonia, hydrogen sulphide, hydrogen chloride, carbon dioxide and argon.

DECOMPOSITION OF LABELLED COMPOUNDS

Unless very high specific activities are used, compounds labelled with radioactive atoms are unlikely to affect the course of chemical experiments in which they are used as tracers. Exceptions can occur when polymerization is possible, or when macromolecules are involved. Similarly, in biological experiments, genetic damage can be multiplied in reproduction.

Radiolytic decomposition of labelled compounds can be a troublesome phenomenon during storage. The wreckage of a molecule after an atom in it disintegrates is not usually significant. Of much greater consequence are the radiation chemical effects of the emitted particles and recoiling fragments. This can be minimized by spreading the active material as a thin layer, or by dilution with unlabelled compound or with some inert substance that can be easily separated. Sometimes, storage on paper can be an advantage. In all these cases the radiation energy is not absorbed in the remainder of the labelled compounds.

Sometimes it is an advantage to exclude oxygen, e.g. to store the compound in a vacuum, or under a nitrogen atmosphere. Low-temperature storage can sometimes minimize decomposition by the emitted radiation. Storage in aqueous solution is usually to be avoided, because of the possibility of decomposition occurring by indirect action. An acid solution, with alcohol added to scavenge the hydroxyl radicals and hydrogen atoms, is probably the best form of aqueous solution. A scavenger, such as benzene, is one of the best diluents to use.

FURTHER READING

† BAYLEY, R. J. and EVANS, E. A., *Storage and Stability of Compounds Labelled with Radioisotopes*, Review 7, Radiochemical Centre, U.K.A.E.A., Amersham, England (1968).

BISHAY, A. (Ed.), *Interactions of Radiation with Solids*, Plenum (1967).

COEKELBERGS, R., GRUCQ, A. and FRENNET, A., Radiation Catalysis, *Advances in Catalysis*, Academic, New York, Vol. 13, p. 55 (1962).

DIENES, G. J. (Ed.), *Studies in Radiation Effects*, Series A: Physical and Chemical, Gordon and Breach, New York, Vol. 1 (1966) and subsequent volumes.

DI MARTINI, R. G. and HUANG, S. R., Radiation Effects in Inorganic Solids, in *Advances in Nuclear Science and Technology*, Vol. 3, Academic, New York (1966).

KIRCHER, J. F. and BOWMAN, R. E., *Effects of Radiation on Materials and Components*, Chapman and Hall, London (1964).

LARIN, F., *Radiation Effects in Semiconductor Devices*, Wiley, New York (1968).

LELL, E., KREIDL, N. J. and HENSLER, J. R., Radiation Effects in Quartz, Silica and Glasses, in *Progress in Ceramic Science*, BURKE, J. E. (Ed.), Pergamon, Oxford, Vol. 4 (1966).

† POLAK, L. S. and PSHEZHETSKY, S. YA., Heterogeneous Processes in Radiation Catalysis, *Pure and Appl. Chem.*, **10** (No. 4), 441 (1965).

Radiation Effects in Inorganic Solids, *Discussions Faraday Soc.*, No. 31 (1961).

TAYLOR, E. H., The Effect of Ionizing Radiation on Solid Catalysts, in *Advances in Catalysis*, Academic, New York, Vol. 18, p. 111 (1968).

WOLFGANG, R., The Hot Atom Chemistry of Gas-Phase Systems, in *Progress in Reaction Kinetics*, PORTER, G. (Ed.), Pergamon, Oxford, Vol. 3, p. 97 (1965).

† WOLFGANG, R., Chemistry of 'Hot' Atoms, *Sci. Am.*, **214** (Jan.), 82 (1966).

† These references should be read first.

Chapter 11

Further Developments

PULSE RADIOLYSIS DEVELOPMENTS

The technique of pulse radiolysis has now been developed to beyond the nanosecond time scale. The fine structure of a linear accelerator pulse consists of a series of spikes, each of several picoseconds duration. These spikes can be used as very short pulses.

A nanosecond is too long for observation in aqueous solution of the parent positive ion which remains after an electron is removed in an ionizing event. However, it has been possible to observe geminate recombination taking place over several nanoseconds. Some of the hydrated electrons ($G \approx 0.3$) return to the $H^{+}{}_{aq}$ ions and a similar number to the $\cdot OH$ radicals. Both of these result from decomposition of the $H_2O\cdot^{+}$. Geminate recombination then takes place.

In the case of certain organic liquids the original positive ions can be observed. For instance $(\text{aniline})^{+}$ has a half life of ≈ 60 nanoseconds. This means that the reactions of these positive species can be followed directly.

There are many more systems that remain to be studied by pulse radiolysis. Knowledge of the early reactions will help to establish complete patterns of reactions, particularly in organic systems.

Quantitative data has been accumulated on the reaction kinetics of various, transient, radical species. This applies particularly to the hydrated electron, for which rate constants for reaction with hundreds of solutes have been measured. This information will help in the understanding of many chemical reactions far removed from the field of radiation chemistry. Many reactions involve radical species as intermediates and, in selected cases, pulse-radiolysis techniques enable a single well-defined reaction to be studied.

MOLECULAR BEAMS

A serious problem in producing detailed models of radiolysis mechanisms is the complexity of the possible excitation states and levels. Little information about them can be derived from normal radiation-chemistry studies.

Molecular-beam measurements can be used to study the dynamics of molecular collisions—elastic, inelastic and reactive. The properties of the species involved in the reaction can be determined, to give a picture of the energetics on a molecular level.

Most experiments have been made on reactions of the type

$$M + XY \longrightarrow MX + Y \tag{11.1}$$

where M is an alkali metal and XY is a halogen(X) compound, with

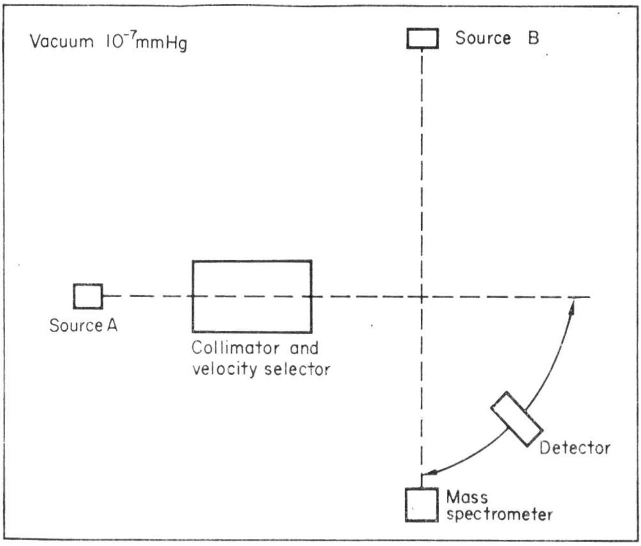

Fig. 11.1 Schematic diagram of molecular beam apparatus.

Y = H, R, X, etc. The study of reactions not involving alkali metals is near, and when this develops it may provide much critical information at a molecular level on radiation-chemistry mechanisms.

The technique consists essentially of two sources producing intersecting, collimated beams, and a detector for measuring the angular dependence of the scattering. Mass spectrometric and velocity analyses of the products can also be made. The apparatus is shown in Fig. 11.1. It must be maintained at a background pressure of about 10^{-7} mmHg (10^{-5} N/m^2).

Results indicate that the reactions proceed through an ion-pair intermediate. This has led to the idea of an 'electron-jump' mechanism, whereby an electron is transferred from the metal atom to the neutral molecule. The transition state activated complex is then formed by coulombic attraction of these oppositely charged species.

$$RCl + Na \longrightarrow RCl^- + Na^+ \longrightarrow [R \cdots Cl^- \cdots Na^+] \longrightarrow R\cdot + NaCl \quad (11.2)$$

Polanyi called this 'harpooning'. It provides further evidence of the importance of ionic intermediates, even in reactions involving radicals.

The duration of the reaction period is less than the average rotational period of the activated complex (5×10^{-13} seconds). Most of the chemical energy appears as internal excitation of the products, primarily as vibrational excitation. Theoretical analysis of molecular-beam interactions is extremely complex, but Monte-Carlo calculations by computer are now producing useful results. In the future, molecular-beam experiments may be of great value in radiation chemistry.

ELECTRON–NUCLEAR DOUBLE RESONANCE

The electron–nuclear double resonance or ENDOR technique is a recent development for resolving e.s.r. hyperfine splittings which normally

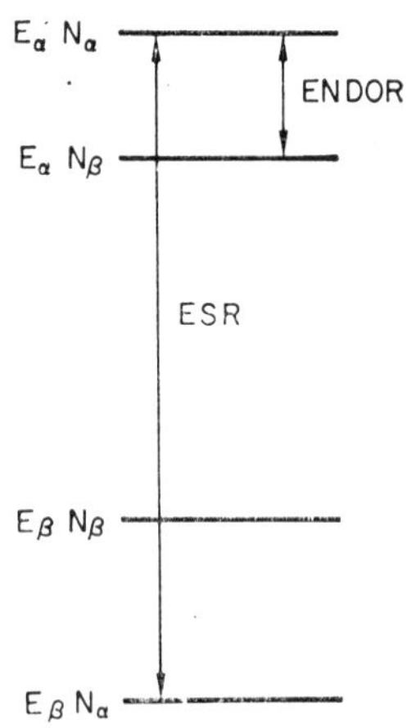

Fig. 11.2 Electron–nuclear double resonance.

overlap. The frequency of a radiofrequency irradiation of the sample is varied while the e.s.r. spectrum is being recorded with a high microwave power level so that the resonance is saturated and the signal is small. At a certain frequency, reversal of the nuclear spins occurs, the population of the upper level of the resonance is altered and the intensity of the e.s.r. signal increases markedly. Figure 11.2 shows a typical set of energy levels with a normal e.s.r. resonance between $E_\beta N_\alpha$ and $E_\alpha N_\alpha$, where α and β

are two different eigenvalues for electron (E) and nuclear (N) spins in the magnetic field. An ENDOR resonance is $E_{\alpha}N_{\alpha}$–$E_{\alpha}N_{\beta}$.

RADIOLYSIS AT HIGH PRESSURES

Irradiations carried out under high pressures of some thousands of atmospheres can reveal much interesting information about reaction mechanisms. The primary yields in the radiolysis of water are not affected by increased pressure, but, where there are competing reactions, that reaction giving products with smaller molecular volumes is favoured.

THEORETICAL CALCULATIONS

A system can be considered to be satisfactorily understood only when a theoretical model can be constructed which will quantitatively predict the response of the system to changes in any combination of parameters. Such theoretical models require a great deal of information on the physical and chemical properties of the system and involve detailed calculations of their interactions. The increasing use of computers will enable the development of more sophisticated models.

In particular, the properties of systems on the microscopic or molecular scale must be known or predicted, as these differ considerably from the bulk properties. For example, the structure of water must be taken into account when considering reactions in aqueous solution. Liquid water does not consist of randomly oriented water molecules, but rather a complex assemblage of interacting entities with considerable structure.

OTHER FIELDS OF CHEMISTRY

The species, reactions and products involved in radiation chemistry are encountered in other fields of chemistry. Hence, knowledge obtained from radiation chemistry studies can be of very general interest. The use of ionizing radiation is one way, and in some cases the only convenient way, of producing a reasonable concentration of certain species and energy states in a form which can be clearly characterized. The reactions of such species can then be followed quantitatively, and this enables a better understanding to be obtained of certain fundamental and simple reactions. In this way, radiation chemistry can make a contribution to general chemical reaction theory.

Solvated electron

The solvated electron is an important chemical reactant. It is the simplest nucleophilic reagent, and the reactions of the solvated electrons, produced during irradiation of suitable polar solvents, with a series of compounds have been investigated. The hydrated electron, produced by irradiation of aqueous solutions, has been studied most. Hydrated electrons may also be made by ultraviolet photolysis of a series of anions, by mixing an amine solution of an alkali metal with water, by passing hydrogen atoms into alkaline solution, by reaction between sodium amalgam and water or at a cathode in aqueous solution. The electrons produced by these means have the same optical absorption spectrum and react in the same way as the radiolytic hydrated electrons, for example with solutes to give adducts.

The discovery of the hydrated electron generated by ionizing radiation and the elucidation of its properties has caused several theories of chemical reaction mechanism to be rewritten.

Hydroxyl radical

The reactions of the hydroxyl radical are also important. It has been claimed that this radical is produced when hydrogen peroxide reacts with certain metal ions, e.g. ferrous ion (Fenton's reagent), titanous ion, etc. In some of these reactions, complexes of the metal ions are thought to be formed, and these complexes rather than the •OH radical may be the reactive species. Sometimes organic peroxides or hydroperoxides are known to be formed, and this complicates the picture further. Although the reactions observed are similar to those of the radiolytic hydroxyl radical, in no case have they been shown to give an identical distribution of isomeric products and the same ratios of relative rate constants for reaction with a series of substituted benzenes. On the other hand, the •OH radical can be produced by photolysis of H_2O_2, and gives the same sort of products as the species formed in radiolysis. Hydroxylation reactions in living organisms are brought about by enzymes, and these enzymatic reactions resemble those brought about by chemical reagents.

Radiolysis has an advantage over other methods of generating hydroxyl radical in that the amount of this species liberated in the solution is known almost as surely as if it were added from a burette. This amount is reasonably independent of varying conditions.

Other species

Hydrogen atoms can be made by passing hydrogen gas over a hot filament or through an electrodeless discharge. On passing the gas stream into aqueous solutions the reactions with solutes have been shown to be identical with those undergone by radiolytic H• atoms.

Some ion–molecule and other reactions involving excited states can be conveniently studied by using ionizing radiation. Several new transient species, such as singly charged ions of divalent metals, e.g. Zn^+, can be prepared and their reactions followed.

The reactions of radicals with various chemical reagents are important not only from the point of view of fundamental chemistry but also for an understanding of many industrial chemical processes. Ionizing radiations are an excellent source of such radicals.

Effect of phase

Ionizing radiation produces much the same yield of primary species in a system, irrespective of the state—solid, liquid or gas—or of the presence of minor components. There are minor differences between phases due to different excitation levels. Differences in reactions and the products must therefore result from differences in the subsequent reactions. Such studies can add greatly to our knowledge of the effect of phase on reaction kinetics.

Vacuum ultraviolet

Radiation at vacuum ultraviolet wavelengths provides a link between ultraviolet and ionizing radiation. Selective photodissociation and, in some cases, photoionization can lead to the formation of some of the species produced by ionizing radiation. This is invaluable for the study of certain chemical reaction mechanisms.

Electrical discharges and plasma

A field in which knowledge obtained from radiation-chemistry studies has been used is the chemistry of reactions in electric discharges or plasma jets. If a sufficiently high potential is applied across a gas it becomes electrically conducting as a result of ionization and electric discharge occurs. A similar state may also be achieved by a suitable radiofrequency input. The ionized and excited gas constitutes the plasma, which can be confined or controlled by an external magnetic field. Ionization and excitation equivalent to plasma temperatures of up to 100 000°K have been obtained.

Reactions of these ions and excited species proceed as in radiation chemistry. There is increasing interest in the possibilities of using plasmas for chemical synthesis, especially of compounds which are thermodynamically unstable at normal temperatures or can only be produced through highly energetic intermediates. Residence times in the plasma jet are normally about 10 milliseconds and reaction rates are very high. Experiments with mixtures of CH_4, NH_3, H_2O and H_2 have produced various amino acids, leading to speculation about the origin of life on earth.

Astrophysics

An unusual application of radiation chemistry is in astrophysics—the study of the formation and properties of the universe. Ions and radicals are produced in space, and their reactions under various conditions can account for many of the phenomena observed.

Ultrasonics

Ultrasonic generators of moderate power are now readily available, being used mainly for cleaning and degradation. These may be primarily physical processes, but chemical reactions, similar to those resulting from irradiation, also occur. The ultrasonic energy produces cavitation in fluids, and this causes bond scission. Subsequent reactions of the fragmentary chemical species are analogous to those in radiolysis.

RADIATION PRESERVATION AND STERILIZATION

There has been an increasing expenditure of effort on the possibilities of using radiation to inhibit the growth of micro-organisms in biological substances, or materials used in biological environments. The major problem is to achieve sufficient inhibition without producing an unacceptable deterioration in properties. However, some successful applications have already resulted, and further developments will certainly follow.

Radiation preservation of food

One field which has been intensively investigated is the preservation of food.

Spoilage of food during the period between production and consumption is both wasteful and expensive. Marked reduction in this spoilage has been achieved for many foodstuffs by irradiation. It is expected that radiation will find a role in the food industry along with the other well-accepted processes of freezing, dehydrating, etc. However, some of the chemical effects of radiation produce flavours and aromas or changes in texture which make the treated material unacceptable. There is also the possibility that some of the chemical products of the irradiation may be harmful toxins or carcinogens.

Radiation effects in foodstuffs can be considered from a number of aspects.

Micro-organisms, such as bacteria and fungi, in food can be killed or reduced in number and potency, so radiation will increase the shelf life of the food. For example, the shelf life of fish, shellfish and meat can be increased 2–5 times, i.e. up to about 30 days, by 0.2 Mrad of radiation. Radiation-preserved bacon has been extensively investigated in the United States.

Physiological and biochemical changes, i.e. metabolism, which proceeds in 'living' food, such as fresh fruit and vegetables, can be affected. Thus, sprouting in potatoes is inhibited, and ripening of bananas and paw paws is delayed. This will increase the effective shelf life of the food. There is the possibility that in killing micro-organisms one also destroys the life processes in 'living' food and thereby decreases the shelf life. This frequently limits the radiation dose to less than that required to control many micro-organisms. The problem may be reduced in some cases by using electron irradiation of low penetration. This treats the skin, which

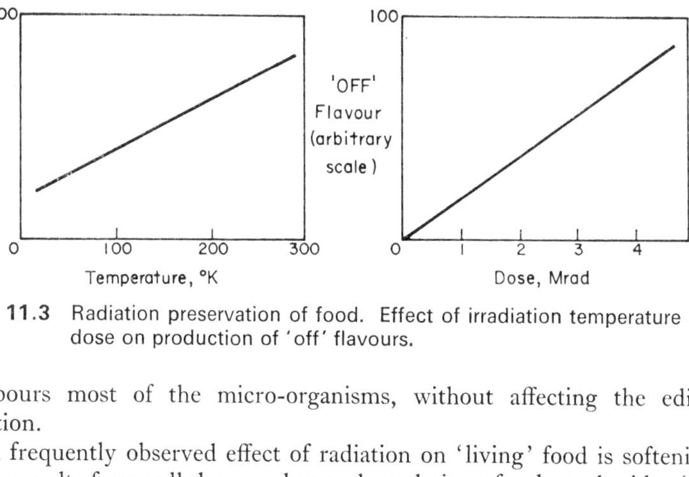

Fig. 11.3 Radiation preservation of food. Effect of irradiation temperature and dose on production of 'off' flavours.

harbours most of the micro-organisms, without affecting the edible portion.

A frequently observed effect of radiation on 'living' food is softening. This results from cell damage due to degradation of polysaccharides (e.g. pectin and cellulose).

Chemical reactions induced by radiation produce many volatile compounds. There is always a great variety of flavour components in any food, and analysis of 'off' flavours is extremely complicated. However, the main causes of unwanted flavours are probably sulphur- and nitrogen-containing compounds resulting from the breakdown of proteins. These chemical reactions are usually reduced when the food is irradiated at low temperatures (Fig. 11.3). Consequently, this may provide a method for extending radiation preservation to a greater variety of foodstuffs. Obviously, decreasing the radiation dose will reduce production of off-flavours and cell damage (Fig. 11.3), but it must be sufficient for preservation.

It is essential that the nutritional value of the food is not appreciably reduced, e.g. radiation has been found to destroy vitamins.

Disinfestation dose levels—those required to sterilize (reproductively) or to kill insects—are generally much lower than those required for the control of micro-organisms and do not usually produce deleterious radiation chemical effects.

Radioactivity produced in the food is negligible for β- and γ-radiation of energy less than 10 MeV (1.6 pJ).

Surgical and pharmaceutical materials

Considerable success has also been achieved in the medical field. Surgical dressings, gloves, sutures and catheters are now routinely sterilized by irradiation. In these cases the reduction in damage to the materials, compared to conventional heat sterilization, is a great advantage. In particular, a wide range of plastics can now be used. One application of this type has been the sterilization of replacement parts for the human body, e.g. heart valves and bone grafts, for which it is the best known method. Pharmaceuticals can also be sterilized by irradiation, and this has proved especially useful for products such as eyedrops and antibiotics for use on the skin. Items such as hypodermic syringes and needles are sealed in plastic bags before irradiation. After irradiation the contents remain sterile until the bag is opened. Household detergents containing enzymes may be sterilized by radiation.

INDUSTRIAL CHEMICAL PROCESSES

At the time of the first development of nuclear reactors, 1945–50, there was great enthusiasm concerning the possibilities of using chemical radiation processes in industrial technology. When it became evident that radiation was not the 'universal catalyst', interest dwindled. Since then, it has become apparent that irradiation can be a superior technique for some processes when all factors are considered—starting materials, ease of operation and control, purity of product, physical properties of product and overall economic considerations. Radiation must now be considered along with other methods for use in some industrial processes.

We will give a brief survey of some applications, ranging from commercial production on a small to moderate scale down to processes which are in the development stage. It should be noted that commercial secrecy surrounds many technological processes, so that not only are full details not available but the fact that radiation is being used is frequently not announced.

HALOGENATION A process is in commercial operation for the direct hydrobromination of ethylene to ethyl bromide

$$CH_2{=}CH_2 + HBr \rightsquigarrow CH_3{-}CH_2Br \qquad (11.3)$$

It proceeds by a chain reaction, hence there is a high radiation chemical yield (G ≈ 20 000).

SULPHOXIDATION Linear-alkyl sulphonates are preferable to aryl, branched-alkyl sulphonates for detergent use, because of their greater bio-degradability. A radiation-induced process for the sulphoxidation of alkanes,

$$RH + SO_2 + \tfrac{1}{2}O_2 \rightsquigarrow RSO_3H \qquad (11.4)$$

was developed in the United States, and promised to be more economic than the conventional method, which uses olefins as feed material. However, a non-radiation process to produce an aryl, linear-alkyl sulphonate was adopted. This product is not as bio-degradable as a linear-alkyl sulphonate, but was considered adequate. There is a plant in Russia producing synthetic detergent by a radiation process.

POLYMERIZATION Radiation is useful for initiating polymerizations at low temperatures. Low temperatures may be necessary to avoid monomer thermal degradation, or to obtain a desirable steric structure. Similarly, polymerization in the solid-state may enhance certain desirable properties. Poly(oxymethylene) of high crystallinity and with a molecular-weight distribution which gives good physical properties has been produced in this way from trioxane. The radiation polymerization of ethylene on an industrial scale is being developed.

CROSSLINKING OF POLYMERS Radiation-induced crosslinking of poly(ethylene) and poly(vinyl chloride) has been developed on a moderate industrial scale. Crosslinking increases the temperature resistance of such materials as the electrical insulation of cables. Crosslinking also introduces a memory effect in poly(ethylene), which is used in heat-shrinkable tube and film.

CONTROLLED DEGRADATION OF POLYMERS Degradation is a process used in industry to lower the average molecular weight of some polymers. The molecular-weight distribution will also tend towards the most probable or normal distribution, which may be much narrower than that of the original product. Controlled degradation may be achieved in suitable cases (polymers with high *G*(scission) values) by irradiation.

SURFACE COATINGS Suitable paints (solvent free) can be cured (polymerized) very rapidly by irradiation. This would permit very rapid throughput—always a desirable feature in industry. Curing ovens or large areas of expensive dust-free factory space are unnecessary, and this results in economic savings. Scanning electron beams and extended source heads are being used for coatings on wood and metal, particularly in the housing, decorating and automobile industries. Special, high-intensity, electron sources are being developed for flash-radiation curing of coatings without harming the material underneath. Since the radiation has to penetrate only a short distance, the voltage provided by conventional power transformers (100–300 kV) is adequate.

WOOD IMPREGNATION The porous structure of wood can be impregnated with a monomer, which will polymerize upon irradiation. This produces a harder, stronger, dimensionally stable material, overcoming a serious defect in the natural product. The process may be difficult, usually involving preliminary evacuation of the wood with heating, and some timbers are unsuitable. Consequently the product is expensive and has been limited to speciality applications.

GRAFTING Poly(olefins) have many desirable properties, such as chemical inertness and ease of fabrication, but they have very poor adhesion, and cannot be printed. A method of overcoming this problem is to graft-polymerize a polar monomer on to the poly(olefin). Acrylic and methacrylic acids have been used successfully for this purpose.

FIBRE MODIFICATION The properties of natural fibres, such as wool and cotton, can be improved by grafting them with synthetic monomers, a process conveniently accomplished with radiation. For example, styrene grafted to wool can improve its anti-creasing and anti-soiling properties.

FIXATION OF NITROGEN There is an enormous market for oxides of nitrogen in the form of nitrate fertilizers. The possibility of using high-energy ionizing radiation to convert air into nitrogen oxides has been investigated. The overall reactions are

$$N_2 + O_2 \longrightarrow 2NO \qquad (11.5)$$

$$2NO + O_2 \longrightarrow 2NO_2 \qquad (11.6)$$

The NO_2 is readily converted to nitric acid, and hence into a form in which the nitrogen is readily available to plant life. The overall reactions are comprised of a variety of reactions involving intermediates derived from both oxygen and nitrogen. The system is a very complex one indeed. The overall yield of fixed nitrogen is $G \approx 6$.

OTHER PROCESSES Other processes have been taken to an advanced stage, but not become marketable propositions due to factors other than technical. In other cases companies have not disclosed that they are marketing products or using intermediates which have been prepared by processes using radiation. The inference is that radiation provides an economic advantage which the companies concerned are not willing to share with their competitors.

Some industrial companies have had active radiation research and development groups for many years. These groups using radiation have shown how products can be made. The final processes chosen for plant-scale production often do not involve radiation, but nevertheless the radiation chemist has played a worthwhile role in showing the potentialities of the product. Being a clean source of radicals, radiation can be used to investigate existing processes, or certain reactions, in order to provide the information necessary to effect improvements. On the reagent shelf alongside the chemical catalyst and the ultraviolet lamp, one must place the radiation source.

CONCLUSION

The processes above indicate the diverse applications of radiation in industry. These, and many more, will undoubtedly lead to radiation becoming a significant industrial technique.

The advances being made in fundamental, academic, radiation-chemistry studies will result in more potential processes. Coupled with the greater understanding of general chemical, biological and related fields, this makes radiation chemistry an interesting and worthwhile subject to study.

FURTHER READING

BADDOUR, R. F. and TIMMINS, R. S. (Eds.) *The Application of Plasmas to Chemical Processing*, Pergamon, Oxford (1968).

Chemical and Food Applications of Radiation Nuclear Engineering, Part 19, Amer. Inst. Chem. Eng. (1967).

† DESROSIER, N. W. and ROSENSTOCK, H. M., *Radiation Technology in Food, Agriculture and Biology*, Avi Publishing Co., Westport, Conn. (1960).

† JEFFERSON, S. (Ed.), *Massive Radiation Techniques*, Newnes, London (1964).

KANAAN, A. S. and MARGRAVE, J. L., Chemical Reactions in Electric Discharges, in *Advances in Inorganic Chemistry and Radiochemistry*, Vol. 6, Academic, New York (1964).

† *Radiation Chemistry and its Applications, Technical Reports Series* No. 84, International Atomic Energy Agency, Vienna (1968).

Radiation Preservation of Foods, Advances in Chemistry Series, No. 65, American Chemical Society (1967).

RAMSAY, N. F., *Molecular Beams*, Oxford Univ. Press, London (1956).

ROSS, J. (Ed.), Molecular Beams, *Advances in Chemical Physics*, Interscience, New York, Vol. 10 (1966).

† These references should be read first.

Appendix 1

Definitions, Units and Conversion Factors

RAD The unit of absorbed dose, corresponds to an energy absorption of 100 erg (10 μJ) per gramme of material irradiated (10^{-2} J/kg). It may be abbreviated to rd. The dimensions are $(\text{length})^2 \times (\text{time})^{-2}$.

ROENTGEN (R) The unit of radiation exposure, is the amount of incident radiation which produces a total of 2.58×10^{-4} coulomb of electrical charge on all the ions of one sign in one kilogramme of air. This is identical with the previous definition of one e.s.u. of charge or 2.08×10^9 ion pairs per 0.001293 g of air (1 cm^3 at 0°C and 760 mmHg pressure). The dimensions are $(\text{electric current}) \times (\text{time}) \times (\text{mass})^{-1}$.

G VALUE (radiation chemical yield) The number of molecules of material changed, or of product formed, for each 100 electron volts (16 aJ) of radiation energy absorbed by the system.

W VALUE The average energy expended in a gas per ion pair formed—usually expressed as electron volts.

M/N VALUE (ion pair yield) The number of molecules of material changed, or of product formed, for each ion pair found in the gas (= G value × W value in eV/100).

LINEAR ENERGY TRANSFER (L.E.T.) The average energy locally imparted to the medium by ionizing radiation of specified energy in traversing a certain distance. It is expressed as energy per unit distance, e.g. eV/Å or aJ/nm, and is different for each type of radiation. (1 eV/Å = 1·6 aJ/nm).

CURIE The activity of a radioactive nuclide is expressed in terms of the number of nuclear transformations which occur in unit time. The unit of activity is the curie (Ci) corresponding to 3.7×10^{10} events, or disintegrations, per second. (s^{-1} = Hz)

HALF-LIFE ($t_{\frac{1}{2}}$) The half-life of a radioactive nuclide is the time taken for the activity to decay to a value of one-half. Since radioactive decay is a first-order process, it has the same value throughout the decay.

Some units

The electron volt cannot be defined exactly in terms of the joule, because it depends on the value of e, the charge on the proton, which is an experimentally determined quantity and subject to revision in the light of new measurements.

$$1\ \text{eV} \approx 0.16021\ \text{aJ}$$
$$1\ \text{keV} \approx 0{\cdot}16021\ \text{fJ}$$
$$1\ \text{MeV} \approx 0{\cdot}16021\ \text{pJ}$$

Factor	Prefix	Symbol
10^{12}	giga	G
10^{9}	tera	T
10^{6}	mega	M
10^{3}	kilo	k
10^{-1}	deci	d
10^{-2}	centi	c
10^{-3}	milli	m
10^{-6}	micro	μ
10^{-9}	nano	n
10^{-12}	pico	p
10^{-15}	femto	f
10^{-18}	atto	a

$$1\ \text{Å} = 0.1\ \text{nm}$$
$$1\ \mu = 1\ \mu\text{m}$$
$$1\ \text{m}\mu = 1\ \text{nm}$$

$$1\ \text{atm} = 760\ \text{mmHg} = 760\ \text{torr} = 101\ 325\ \text{N/m}^2 = 1.013\ 25\ \text{bar}$$

A molar solution (1M) = 1 mol/litre = 1 mol dm^{-3}.

The units for second order kinetic reaction rate constants are:—

$$1\ \text{M}^{-1}\text{s}^{-1} = 1\ \text{litre mol}^{-1}\text{s}^{-1} = 1\ \text{dm}^3\ \text{mol}^{-1}\ \text{s}^{-1} = 10^3\ \text{cm}^3\ \text{mol}^{-1}\ \text{s}^{-1} = 10^{-3}\ \text{m}^3\ \text{mol}^{-3}\ \text{s}^{-1} \approx 1\ \text{kg mol}^{-1}\ \text{s}^{-1}$$

Molar decadic absorptivity (molar decadic absorption or extinction coefficient)

$$1\ \text{litre mol}^{-1}\ \text{cm}^{-1} = 1\ \text{M}^{-1}\ \text{cm}^{-1} = 0{\cdot}1\ \text{m}^2\ \text{mol}^{-1}$$

Conversion factors

1 rad	= 100 erg absorbed/g
	= 10 μJ/g
	= 0.01 J/kg
	= 6.24×10^{13} eV/g
	= 10^{-5} watt second/g (Ws/g)
	= 2.4×10^{-6} cal/g
	= 1.83×10^{12} ion pairs/g of air.
1 watt-minute	= 6 gramme Mrad
1 eV per molecule	= 23.06 kcal per mole
	= 96.49 kJ per mole
1 pile unit	= 16–80 Mrad
1 cal	= 4.184 joule
1 erg	= 0.1 μJ
0° C	= 273.15 K
1 gauss	= 2.80 MHz (free electron e.s.r. resonance position)
Density of air	= 0.001293 gramme per cm^3 (0°C, 760 mmHg)
Velocity of light (c)	= 2.998×10^8 metre per second
Charge on electron	= 1.602×10^{-19} coulomb
Avogadro's constant (L, N, N_A)	= 6.0225×10^{23} molecules per mole
Gas Constant (R)	= 8.314 J mol^{-1} K^{-1}
Thermal Energy (at 20°C)	= 0.025 eV per species
	= 2.4 kJ per mole
	≈ 600 cal per mole

We have found it worthwhile to commit to memory the following most useful conversions:

1 watt-minute = 6 gramme Mrad

1 Mrad changes G × 1.036 × 10^{-6} mole per gramme irradiated

or in dilute aqueous solution G × 1.036 mmol/litre.

This relationship can be readily derived as follows

$$1 \text{ rad} = 100 \text{ erg/g } (10^{-2} \text{ J/kg})$$
$$= 6.242 \times 10^{13} \text{ eV/g}$$

and converts $6.242 \times 10^{13} \times G/100$ molecules/g

or $$6.242 \times 10^{11} \times G/N_A = \frac{6.242 \times 10^{11} \times G}{6.023 \times 10^{23}} \text{ mol/g}$$
$$= 1.036 \times G \times 10^{-12} \text{ mol/g}$$

Hence 1 Mrad converts $G \times 1.036 \times 10^{-6}$ mol/g

or $G \times 1.036 \times 10^{-3}$ mol/kg

Electron irradiation

Electrons of energy 1 MeV, in material of density 1 g/cm^3, penetrate effectively 3.5 mm and have maximum penetration of 4.5 mm.

For β-particles, mean energy is 0.4 × maximum energy.

Photon irradiation

Water exposed to 1 R of 1-MeV γ-radiation receives a dose of 0.965 rad. Air receives a dose of 0.869 rad.

Cobalt-60 emits two γ quanta, with energies of 1.173 and 1.332 MeV, on each disintegration. The specific γ-ray constant (Γ) of ^{60}Co is 3.67×10^{-4} R m^2 s^{-1} Ci^{-1}. At one metre from one curie of ^{60}Co the absorbed dose rate in water is 1.27 rad/h and in air 1.15 rad/h. The exposure rate is 1.32 R/h.

A photon of wave number 8066 cm^{-1} (wavelength 1240 Å = 124 nm) has energy 1 eV.

Relative doses

Relative doses may be calculated from

$$D_2/D_1 = (Z/A)_2/(Z/A)_1$$
$$D = \text{absorbed dose (rad)}$$
$$Z = \text{atomic number}$$
$$A = \text{atomic weight.}$$

For materials of similar composition, the absorbed dose is approximately proportional to the density. This is particularly useful for aqueous solutions.

More accurately, *mass energy–absorption coefficients* or *mean mass stopping power* ratios should be used.

The *mass energy–absorption coefficient* (μ_{en}/ρ) of a material is a measure of the energy absorbed from an incident photon (γ-ray or x-ray) per unit (thickness $\times$ density) of material. Units are cm^2/g ($10^{-1}m^2kg^{-1}$).

The *mass stopping power* (S/ρ) is the energy lost by a charged particle, e.g. electron, of specified energy per unit (thickness $\times$ density) of material. Units are erg cm^2/g ($10^{-8}Jm^2kg^{-1}$).

The attenuation of radiation as it passes through materials can be calculated using tables of these quantities. Relative dose rates of exposed materials can also be calculated. These are some typical values.

Material	(μ_{en}/ρ) cm^2/g 1 MeV γ	(s/ρ) relative to that for air, erg cm^2/g Electrons, 1 MeV	Compton electrons from 1 MeV γ	^{60}Co γ
H	0.056	2.39–2.44	2.44–2.50	
Air, O,N,C	0.028	0.98–1.01	0.98–1.01	1.0
$(CH)_n$	0.030			1.12
$(CH_2)_n$	0.032			1.21
Water	0.031			1.13

Energy absorbed by a thin layer of material (erg/g) for γ- or x-irradiation = mass energy–absorption coefficient (cm^2/g) $\times$ energy incident on layer (erg/cm^2).

Energy loss per unit length (erg/cm) for electrons = mass stopping power (erg cm^2/g) $\times$ density (g/cm^3).

Exact definitions are given in United States National Bureau of Standards Handbook 85, *Physical Aspects of Irradiation* (1962). Units and conversion factors, including the SI system, are covered in detail in MCGLASHAN, M. L., *Physico-Chemical Quantities and Units*, Royal Institute of Chemistry, London (1968).

The International Commission on Radiation Units and Measurements Report 11 (1968) and others in the series are available from ICRU Publications, P.O. Box 4869 Washington, D.C. 20008, U.S.A.

A.S.T.M. (American Society for Testing and Materials) Standard D2568 is a useful guide for calculating the absorbed dose in a substance from the dose measured in a dosimeter.

Appendix 2

Typical Calculations

The following worked example will illustrate how radiation dose and radiation chemical yield can be calculated from experimental data. Reference should be made to *Appendix 1* for information on Units and Conversion Factors.

Dosimetry

10 cm^3 of Fricke dosimeter solution in a flask was exposed for 23.6 minutes at 20 cm from a point source of cobalt-60. The absorbance (optical density) of the irradiated solution was found to be 0.721 at 304 nm when measured in a 1-cm cell at 20°C and after allowing for the cell correction.

Calculate:

(i) the absorbed dose rate in rads per hour at this position;

(ii) the exposure rate in roentgens per hour;

(iii) the strength of the source in curies;

(iv) the dose rate three months later.

Calculation

Decadic absorbance (A) or Optical Density (O.D.)

$= \log_{10}(I_0/I)$, where I_0 is incident light intensity and I is transmitted light intensity,

$=$ molar decadic extinction coefficient (ε) $\times$ concentration (mol/dm^3) $\times$ cell path length (1 cm)

ε at 25°C is 2193 $dm^3\ mol^{-1}\ cm^{-1}$.

The variation of ε with temperature is given by

$$\varepsilon_2 = \varepsilon_1[1 + 0.007\,(t_2 - t_1)]$$

hence $\varepsilon\,(20°C) = 2193\,(1 - 0.007 \times 5) = 2116$

$= A/\text{concn or O.D./concn} = 0.721/\text{concn}$

$\therefore$ Ferric ion concentration $= 0.721/2116 = 0.341 \times 10^{-3}\ mol/dm^3$

1 Mrad or 1000 krad produces change of

$$G \times 1.036 \times 10^{-6} \text{ mol/g}$$

or $$G \times 1.036 \times 1.024 \times 10^{-3} \text{ mol/dm}^3$$

where 1.024 g/cm^3 is density of dosimeter solution.

For Fricke dosimeter $G(Fe^{3+}) = 15.5$.

D krad produces change of 0.341×10^{-3} mol/dm^3.

$$\therefore \quad \text{Dose (D) is } 1000 \times \frac{0.341}{15.5 \times 1.036 \times 1.024} = 20.7 \text{ krad}$$

OR in each cm^3 of solution 0.341×10^{-6} mol of ferric ion is produced, i.e. $0.341 \times 10^{-6} \times N_A$ molecules. (N_A is the Avogadro constant).

Since 100 eV produces 15.5 molecules of Fe^{3+}, dose (D) must be

$$100 \times \frac{(0.341 \times 10^{-6} \times 6.023 \times 10^{23})}{15.5} = 1.32 \times 10^{18} \text{ eV/cm}^3$$
$$= 1.29 \times 10^{18} \text{ eV/g}$$

Since 6.24×10^{13} eV/g = 1 rad

this is $\frac{1.29 \times 10^{18}}{6.24 \times 10^{13}} = 2.07 \times 10^4$ rad or 20.7 krad.

(i) Absorbed dose rate is 20.7/23.6 krad per minute

= 52.7 krad per hour

The dose of 20.7 krad is within the range of validity of the dosimeter. If a series of measurements have to be made, all the constants can be collected into one factor by which each absorbance must be multiplied to give absorbed dose in rads. In this case the factor is

$$10^9/(2116 \times 15.5 \times 1.036 \times 1.024) = 2.87 \times 10^4$$

The volume 10 cm^3 does not enter into the calculation. We deal with concentrations, or with one cm^3, or one gramme. Absorbed dose, in rads or joules per gramme, is a measure of concentration of added energy. Alternatively, radiation can be considered as a reagent, which is added to oxidize the ferrous ion. Dose is then expressed as amount of radiation reagent added per gramme of solution.

Dose to water is $52.7 \times 1.024/1.022$;
or from $\sum$[weight fraction $\times (\Sigma Z/\Sigma A)$], $52.7 \times 0.5556/0.5538$; or more accurately from mass-energy-absorption coefficients, $52.7 \times 0.02975/0.02965 = 52.7/0.997 = 52.9$ krad per hour.

1 roentgen deposits 0.965 rad in water.

(ii) *Exposure rate* is 52 900/0.965 = 54 800 R per hour

1 Ci Co-60 gives 1.32 R per hour at 1 metre

A 54 800 0.2

(iii) *Strength of source* (A) $= \left(\frac{54\,800}{1.32}\right) \times \left(\frac{0.2}{1}\right)^2 = \underline{1660\text{ Ci}}$

OR 1 Ci Co-60 deposits 1.27 rad per hour in water at 1 metre

A 52 900 0.2

$$A = \left(\frac{52\,900}{1.27}\right) \times \left(\frac{0.2}{1}\right)^2 = \underline{1660\text{ Ci}}$$

Cobalt-60 loses 3.3% of its activity in 3 months.

$$96.7\% \text{ of } 52.9 = \underline{51.2\text{ krad per hour}}$$

OR, more accurately, using the law of radioactive decay $N_t = N_0 e^{-\lambda t}$ and knowing that $\lambda = 0.01096$ if t is in months,

$$\log N_t = \log N_o - \frac{0{\cdot}01096}{2{\cdot}303} \times 3$$

$$= 1.724 - 0.014$$

$$N_t = 51.3$$

(iv) Dose rate in water after three months will be $\underline{51.3\text{ krad per hour.}}$

In practice, the dose rate would also be redetermined by Fricke dosimetry at the time of the next measurement.

Irradiation

10 cm^3 of cyclohexane (density 0.78 g/cm^3) was then exposed at the same position for 4.5 hours. 0.25 cm^3 (corrected to 0°C and 760 mmHg pressure) of hydrogen was produced by radiolysis.
Calculate:

(v) the radiation chemical yield of hydrogen;

(vi) the percentage of the cyclohexane evolved as hydrogen.

Calculation

Firstly, the ratio of the absorbed dose rate in the cyclohexane to that in water must be calculated.

(*a*) assuming 1 g of cyclohexane receives same dose as 1 g of water

(*b*) from Z/A where Z is the atomic number and A is the mass number

$$\text{factor is} \quad \frac{(\Sigma Z/\Sigma A) \text{ for } C_6H_{12}}{(\Sigma Z/\Sigma A) \text{ for } H_2O} = \frac{48/84}{10/18} = 1.03$$

(*c*) from mass-energy-absorption coefficients *

* Tables and references are given in *U.S. NBS Handbook* 85 (1962).
Mass–energy absorption coefficient (μ_{en}/ρ), cm^2/g.

Photon energy (MeV)	1.0	1.5
H	0.0555	0.0507
C, N, O	0.0280	0.0255

using elemental composition

$$\frac{(6C + 12H)/\Sigma A}{(2H + O)/\Sigma A} = \frac{(6 \times 12 \times 0.0267 + 12 \times 0.0531)/84}{(2 \times 0.0531 + 16 \times 0.0267)/18} = 1.03$$

OR using listed values *

$$\frac{(\mu_{en}/\rho) \text{ for } (CH_2)_n}{(\mu_{en}/\rho) \text{ for water}} = \frac{0.0305}{0.0297} = 1.03$$

This illustrates the points one must consider. (*a*) does not take account of elements present, (*c*) is the value that should be used. If data is not available, then Z/A, i.e. (*b*), may be used when Compton scattering processes predominate. In this case it happens to be correct.

$$\text{Dose rate } 1.03 \times 51.3 = 52.7 \text{ krad per h}$$

4.5 h irradiation gave 0.25 cm^3 H_2 from 7.8 g of cyclohexane, i.e.

$$52.7 \times 4.5 = 237 \text{ krad gave } \frac{0.25}{7.8 \times 22\,400} \text{ mol from 1 g}$$

$$1000 \text{ krad gives} \quad G \times 1.036 \times 10^{-6} \text{ mol/g}$$

$$G = \frac{0.25}{7.8 \times 22\,400} \times \frac{1000}{237} \times \frac{1}{1.036 \times 10^{-6}}$$

(v) $\underline{G(H_2) = 5.8}$

$$\frac{0.25}{7.8 \times 22\,400} \text{ mol } H_2 \text{ weighs } \frac{2 \times 0.25}{7.8 \times 22\,400} \text{ g}$$

(vi) percentage of cyclohexane is

$$\underline{2.9 \times 10^{-4}\ \% \text{ by wt}}$$

1 gramme of cyclohexane is 1/84 mol

$$\frac{0.25}{7.8 \times 22\,400} \times \frac{100}{(1/84)} = 1.2 \times 10^{-2}\ \% \text{ mol/mol}$$

If G(−cyclohexane), the radiation chemical yield of cyclohexane decomposition, were known it would be possible to calculate the total percentage decomposition. If more than 5–10% of target material is consumed the G value might be no longer linear with dose.

When investigating reaction mechanisms it is advisable to use a dose which converts less than 1% of the original material provided this gives sufficient product for an accurate analysis. At higher doses attack on the products becomes progressively more important and makes quantitative interpretation more difficult.

Self-decomposition

Cyclohexane labelled with radioactive carbon-14 is listed in the catalogue of the Radiochemical Centre, Amersham, as *cyclo*-Hexane-C14(U) 2–10 mCi/mM. Assuming the G value calculated above and a specific activity of 10 mCi/mmol, calculate for one gramme and for one month, (vii) the volume of hydrogen gas at 20°C and 760 mmHg pressure and (viii) the percentage of the total hydrogen, evolved as gas by self-radiolysis.

Calculation

1 g cyclohexane is $\frac{1}{84}$ mol, i.e. $\frac{10}{84}$ Ci

In one month this undergoes

$$\frac{10}{84} \times 3.7 \times 10^{10} \times 3600 \times 24 \times 30 \text{ disintegrations}$$

each of maximum energy 156 000 eV and $E_{mean} = 0.4 \times E_{max}$

100 eV gives G (= 5.8) molecules OR $5.8/N_A$ mol of hydrogen

Hydrogen produced is

$$\frac{3.7 \times 10^{11} \times 3600 \times 24 \times 30}{84} \times \frac{0.4 \times 156\,000}{100} \times \frac{5.8}{6.023 \times 10^{23}}$$

$$= 6.9 \times 10^{-6} \text{ mol}$$

$$= 6.9 \times 10^{-6} \times 22\,400 \times \frac{293}{273} \text{ cm}^3 \text{ at } 20°\text{C}$$

(vii) $= \underline{0.17 \text{ cm}^3 \text{ from each g cyclohexane}}$

Total hydrogen present is $\frac{12}{84} = \frac{1}{7}$ wt of cyclohexane

$$= \frac{1}{14} \text{ mol of } H_2 \text{ per g of cyclohexane.}$$

$$\therefore \text{ percentage is } \frac{6.9 \times 10^{-6} \text{ mol evolved per g}}{1/14 \text{ mol present per g}} \times 100$$

(viii) $= \underline{9.7 \times 10^{-3}\ \% \text{ of total hydrogen.}}$

If G values are known, other decomposition products, or the total decomposition, can be evaluated.

Shielding

The cobalt source is placed at the centre of a cubic box of side 60 cm. Calculate (ix) the thickness of lead required to line the box so that the exposure rate on the outside at the centre of each side is 10^{-3} R per hour at 5 cm from the face. (hvt for lead is 1.06 cm.)

Calculation

Unshielded 1 Ci cobalt-60 exposure rate 1.32 R per hour at 1 m

1660 X 0.35

$$\text{Exposure per h } (X) = 1.32 \times \frac{1660}{1} \times \left(\frac{1}{0.35}\right)^2 = 18\,000 \text{ R}$$

$\therefore$ must reduce 18×10^6 fold

i.e. 2^{24} fold

Half-value thickness of lead is 1.06 cm $\therefore$ will require 24×1.06 cm (ix) of lead = 25.5 cm.

The same answer is obtained by using the exponential attenuation law:

$$I = I_0 e^{-\mu x}$$

where I_0 is the intensity of the unattenuated radiation, I is the intensity of the attenuated radiation, x is the thickness of the attenuating medium and μ is its attenuation constant.

Then,

$$I_0/I = e^{\mu x}$$
$$\ln (I_0/I) = \mu x$$
$$\log (I_0/I) = \mu x/2.3$$
$$\log (I_0/\tfrac{1}{2}I_0) = \mu x_{\frac{1}{2}}/2.3$$
$$\log (2) = 1.06\,\mu/2.3$$
$$\therefore \quad \mu = 2.3 \log (2)/1.06 = 0.693/1.06 = 0.654 \text{ cm}^{-1}$$

and

$$x = 2.3 \log (I_0/I)/\mu$$
$$= 2.3 \log (18\,000/10^{-3})/0.654$$
$$= (2.3)\,(7.255)/0.654$$
$$= 25.5 \text{ cm}$$

Index

absolute rate constants 60
absorbed dose 20, 160
absorbed energy 20
absorption coefficients 164, 166, 168
absorption of radiation 8
  electrons 12
  fission fragments 14
  heavy particles 14
  neutrons 14
  x-rays and γ-rays 10
accelerators **34**, 58
alcohols 102
alkanes 103
alkyl halides 105
alpha-emitters 34
alpha-particles 8, **14**
amino acids 133
annealing 57, 142
aqueous systems 61, **80**, 151
  benzene 97
  ethyl alcohol 96
  ferrous ion 95
Argonne syringe technique 47, Plate 3
aromatic compounds 109
astrophysics 154
Auger effect 11, 19

Becquerel 3
benzene 109
  in aqueous solution 96
beta-particles 8, **12**, 163
bibliographies 5
biochemicals 128, **131**
biological
  materials 128
  effects 2
  macromolecules 126

blobs 17
branch track 17
Bremstrahlung 13
build-up 12

caesium-137 34
cage effect 24, 68, 101, 107
  in polymers 121
calculations 62, 151
calorimetry 38
carbohydrates 132
cascade 13
catalysis 143
cave facility 33
cells, biological 129
Cerenkov radiation **13**, 59
ceric dosimeter 41
chain reactions 72, 76, 78, 106, 108, **113**, 124
characteristic x-rays 10
charge transfer 67, 69, 72, 90, 92–3, 105, 143
charge trapping 142
chemonuclear processes 37
chlorates 140
clusters 65, 75
cobalt-60 31, 163
coherent scattering 11
competition kinetics, **49**, 93, 107
Compton scattering 10, 16
concentrated solutions 94
conferences 6
copolymerization 115, 117
  graft **124**, 158, Plate 8
Coulombic interaction 13
crosslinking of polymers **121**, 157, Plate 7
cube-root relationship 82

curie 160
Curie, Madame 3
cyclotrons 36

Debierne 3
defects 138
degradation of polymers **119**, 157, Plate 6
definitions 160
delta-ray 16
detergents 156–7
deuterons 14
dielectric constant 22, 63, 100
diffusion 23, 63
direct effect 28, 94, 131
disinfestation 155
displacement (*see* knock-on)
dissociative electron capture **68**, 107
DNA 129, 135
dose 20
dosimetry 37
  calculations 165
  calorimetry 38
  ceric 41
  chemical 39
  Fricke (ferrous sulphate) **39**, 60, 95, 165
  internal 39
  ion chamber 38
  nitrous oxide 39, 77
  other dosimeters 41
  of pulses 60
  solid state 41
dose-rate effects 27
dry electron 24

electrical discharge 153
electric field 38, 66, 70, 72, 105
electrode processes 144, 152
electromagnetic radiation 9
electron beams 8, **12**, 163
electron
  capture 24, 68
  escape 22, 99
  transfer (*see* charge transfer)
  trapping 24, 84, 100
electron spin resonance 53, Plate 4
  of biological macromolecules 126
  of inorganic solids 139

ENDOR (electron-nuclear double resonance) 150
energy transfer 28, 67, 71
enzymes **134**, 156
e.p.r. (*see* electron spin resonance)
Essex technique 72
e.s.r. (*see* electron spin resonance)
ethanol aqueous solution 96
ethyl bromide 156
excited molecules 3, **15**, 22, 80, 100
  in gases 66, 71
excited ions 16
excited radicals 71
exposure 20, 160, 163

fats 136
F-centre 56, 139, 143
Fenton's reagent 152
ferrous solutions 95
fibre modification 125, **158**
field emission 58
films 7
fission fragments 9, **14**, 37
fixation of nitrogen 158
flow methods 52
food preservation 154
fractionated dose—biological 130
free ions 25, **99**
free radicals (*see* radicals)
freeze–thaw technique 46
Fricke 3
  dosimeter **39**, 60, 95, 165
fuel elements 34
fundamental studies 2

gamma-rays 8, **10**, 163
gases 18, 62, **65**
geminate recombination 22, 148
  in organic compounds 99
glass 141
glasses, organic 56
glassware cleaning 46
graft copolymerization **124**, 158 Plate 8
g value (yield) 25
G value 25, 160

half-life 161
half-value thickness 8

halides 105
hazards 43
heterogeneous catalysis 143
heavy charged particles 18
high pressures 151
history 3, 65
$HO_2\cdot$ (*see* hydroperoxyl radical)
hot
  atoms 144
  radicals 71
  reactions 29
hydrated electron 61, **84**, 152
hydrogen
  abstraction 88, 92, 121, 133, 134
  atoms 88, 101, 152
  halogen reactions 78
  ortho–para conversion 76
hydroperoxyl radical 85, 88, **92**
hydroxyl radical **90**, 152

impurities 44
inactivation 129
indirect effect 28, 131
induced radioactivity 15
industrial processes 2, 37, 125, **156**
industrial catalysis 144
inorganic solids 138
insects 155
interaction of radiation 8
  chemical aspects 21
  physical aspects 15
intermediates 68
intermittent field technique 70
internal sources 39
interrupted doses 51
interstitial defect 138
ionic reactions
  in organic compounds 104
ionization 11, 13, 15, 65, 106, 108
  chamber 38, 66
  potential 66
ion lifetimes 70
ion–molecule reactions 3
  in gases 70
  in organic compounds 101
  in alkanes 104
ion pair yield 38, 160
isolated spurs 17
isotope sources 31
isotopic labelling 74, 146

Kailan 3
kinetics 49
kinetic salt effect 84
knock-on 14, 139

labelled compounds 146
L.E.T. (linear energy transfer) **19**, 23, 68, 102, 160
  in biological systems 131
Lea–Gray–Platzmann model 22
lethal dose 130
Lind 3
light charged particles 12, 16
linear accelerator **35**, 58
lipids 136
liquid nitrogen 43
literature 4 and at end of each chapter
low temperatures 56
luminescence 22

machine sources 34
macromolecules (*see* polymers)
mass spectrometry 68
materials 145
matrix isolation 56
meetings 6
memory effect 123
molecular beams 148
molecular mobility 118
molecular products
  in organic compounds 101
  in water 81, **82**
monomers 62, **113**
monitors, radiation 42
M/N value 38, 160
Mund 3

neutralization 22, 69, 71, 75, 76
neutrons 8, **14**, 36
nitrogen fixation 158
nitrous oxide
  dosimeter 39
  radiolysis 77
  scavenging 73, 85

nuclear chemistry 1
nuclear reactors 2, 15, **36**, 83, 144
nucleic acids 135

OH (*see* hydroxyl radical)
olefins 108
organic compounds 62, 79, **99**
oxidizing radical 83
oxygen **74**, 93
ozone **43**, 75

paints 157
pair production 11
panoramic radiation facility 33
papers 5
particles (*see* radiation)
penetration 8
  electrons 8, 163
  x-rays and γ-rays 9
peptides 133
perbromate 141
peroxides 94, 119, 137
pharmaceuticals 156
phase effects 153
pH effects 25, **82**, 85, 90, 92, **94**, 95–8, 125
photobleaching 56, 142
photochemistry 1, 3
photoelectric effect **11**, 16
plasmas 153
plastics 145
polymerization 113
  anionic 115
  cationic 114
  copolymerization 115, 117
  emulsion 116
  free radical 115
  graft **124**, Plate 8, 158
  industrial 125, **157**
  post-irradiation 118
  solid state 117
polymers 2, 62, 113, **119**
  aromatic 124
  biological 126
  crosslinking **121**, Plate 7, 157
  degradation **119**, Plate 6, 157
  gel formation 122
  indirect action 124
  molecular weight 119, 122
  molecular weight distribution 120
  protection 124
  radiolysis products 123
  scission 119
polypeptides 133
polysaccharides 132
pond facility 33, Plate 2
post-irradiation reactions 25, 118
preservation of food 154
pressure effects 151
primary processes 29
  in aqueous systems 80
primary products 26, 50
primary reactions 69
primary species 23
  in water 81, **83**, 97
protective effect 29 , 95, **110**, 131, 142
proteins 133
protons 8, 14
pulse radiolysis 58, Plate 5, 63
  aqueous systems 61
  developments 148
  gases 62, 79
  monomers and polymers 62
  organic compounds 62
purification 44

quality of radiation 10
quartz 141

rad 20, 160, 162
radiation 8–15
  Cerenkov **13**, 162
  electromagnetic 9
  hazards 43
  reactor 15
  units 20, 162
radiation chemical yield 25, 160
radiation chemistry, definition 1
radiation sources 31
  cave 33
  cyclotrons 36
  fuel elements 34
  internal 34
  isotope 31, Plate 1
  linear accelerator **35**, 58
  machine 34
  plaque 34

radiation sources (*contd.*)
pond 33, Plate 2
reactors 15, **36**, 83, 144
transformers 36, 158
Van de Graaff generators 35, **58**
radicals 3, 21
acid–base properties 85, 94
excited 71
e.s.r. 54
hot 71
in organic compounds 101
trapping 56, 101
radical reactions 21, **23**
in alkanes 104
in aqueous systems 81
in gases 69, 72
in organic compounds 101
radioactivity 3, 10, 15, 37
radiosensitivity 130
range 8
rate constants 21, 161
rates of reaction
of electron 87
of hydrogen atom 89
of hydroxyl radical 91
Rayleigh scattering 11, 19
rays (*see* radiation)
R.B.E. (relative biological effectiveness) 131
reactors (*see* nuclear reactors)
recombination 22–3, 29, 81, 101
reducing species 83
references 4–7 and end of each chapter
relative doses 163, 167
relative rate constants 50
repair phenomena 130
reproductive death 129
reviews 5
Roentgen 3
roentgen **20**, 38, **160**, 163
rotating sector 51

safety 42
sample
storage 46
purification 44
treatment 46
Samuel–Magee model 22
saturation current 66
scanned beam 35
scattering 11
scavengers 25, 49, 101
in gases 73
nitrous oxide 73, 85
scission of polymers 119
sector, rotating 51
secondary products 26
self-irradiation 34, 144, 146, 169
sensitization 130
shielding 42, 170
short track 17
silica 141
solids 19
solid-state polymerization 117
solvated electron 24, 151
sources 31
spatial distribution of events 18
spectra
of electron 61, 84
of hydroxyl radical 90
spurs 17
steady state 28
sterilization 3, 156
stopping power 164
storage 46, 146
strontium-90 34
sub-excitation electrons 16, 67
surface coatings 157
surgical materials 156
survival fraction 129
syringe technique 47, Plate 3

target theory 130
theoretical models 62, 151
third body 68
threshold dose 130
time scale of events 17
topochemical effects 118, 140
topotaxy 118
tracks 17
transfer of energy 67, 71
transient species 58, 148, 152
triplet state 71
triply distilled water 45
tritium 34

ultrasonics 154
ultraviolet 153
units 20, 161

vacuum ultraviolet 153
Van de Graaff generator **35**, 58
V-centre 138, 143

water (*see* aqueous systems)
wood impregnation 158
W-value 20, 66, 160

x-rays 8, 34

yield 25